To G
M

David + Mary.

FROM
LAND TO SEA

FROM
LAND TO SEA

REGINALD SANDERSON

SILENT BOOKS

First published in Great Britain 1990
Silent Books, Swavesey, Cambridge CB4 5RA

© copyright Reginald Sanderson 1990

No part of this book may be reproduced in any form, or by any means, without prior permission in writing from the publisher.

ISBN 1 85183 026 X

Typeset by Goodfellow & Egan Phototypesetting Ltd., Cambridge
Printed in Great Britain by
St Edmundsbury Press, Bury St Edmunds, Suffolk

CHAPTER ONE

'Splash, splash', went the rain. It was the month of April, although I was too young to know that, and watching the rain bubbling upon the road fascinated me. I could only be about five, and this was the first thing in my life that I can remember.

My parents lived in a house on the outskirts of the village. On the opposite side of the road stretched a long belt of trees, and inside these trees was the big Park House, converted into a school for gentlemen's sons. The grounds attached to this park consisted of a well-mown field, where tennis courts stretched from end to end, and a cricket pitch and golf course. Many a happy hour was I to spend inside this park, finding golf balls and selling them back to the golfers for as much as sixpence each – which was then very good money.

This park also held a big meadow where cattle grazed and drank from the pond. Dividing the meadow from the thickest part of the wood was a lovely old path called Primrose Walk, lined with old leafy trees which made ideal nesting places for the blackbird and thrush, with the old crow 'Caw, Cawing' overhead.

Amongst these beautiful surroundings I spent the first years of my life. As I grew up, my ambition was to have a hundred pounds, and a small farm by the time I was twenty-one. I used to tell my aunt this, and she wisely used to shake her head and say 'We will wait and see'.

My parents moved from one end of the village to the other and I now lived by a farm, which lay between two hills, with a watercourse stretching through many fields. In winter this would be flowing with water, and then as children we would hold boat races with pieces of wood or paper boats, but in summer it would dry up and blackbirds and thrushes would build their nests in its mossy banks. Here for the next few years I learned from my family and companions quite a lot about country life. I could hunt the rabbit and hare, and gather every kind of bird's egg by taking one from each nest.

It was here I first went to work, harrowing in the corn with a horse-drawn set of harrows; during the winter months I was given the job of scaring the birds off the fresh-sown corn. This was mostly on Sunday, when every one slept. It was wonderful to have the whole of the countryside to oneself and watch the break of the day as others slept. All day long I would watch over my kingdom which the birds and animals shared with me.

As I grew up, the teaching at school was opening up the world beyond. History and geography books were my main interest, while in all kinds of sport I was eager to give of my best. As I reached school-leaving age, I was an ordinary country boy, strong and healthy but shy.

Now, at the age of fourteen, I was to stand on my own two feet and help my parents with my small income which I received for the long hours and hard work I put in.

The first thing one did when starting work and earning money was to buy a bicycle, as the village was almost two miles away. We had again moved house – this time to another village, an old and pretty country place, its many thatched houses giving it an old-fashioned charm. The largest house, which was very old, was called 'The Temple', getting its name from the Knights Templars who settled there after the Crusades.

On the other side of the village was the village common, with a stream bordering the boundary, deep enough for a boy to bathe in. Clumps of bushes dotted this space, and the villagers by an old law were allowed to graze many head of cattle.

The church and chapel and Memorial Hall were its main meeting places while the leafy lanes were where the pretty maids of the village paraded in their Sunday dresses to capture the hearts of the boys.

At the age of fourteen I was taking lessons in batting as I wanted to get into the second eleven of the village cricket team, and I had made up my mind to play steadily, to get picked to play Saturday afternoon matches, so I jumped on my new racing cycle and I was down on the cricket field almost every night. If the schoolmaster happened to be there, his advice and tips were inwardly digested. I had to train hard as once a year the flower show was held in the Temple grounds, and the men and boys pitted their skill and strength in games of every kind. The village was sparkling with life and excitement that Saturday morning, and to finish it off the village band with its large following of young people would be playing its latest tunes. One and all would join in the singing and dancing and long would sound the laughter echoing with the music through the village. Many friends and relations would meet and have a farewell drink at one of the village pubs, declaring how good it was to meet again. Although people were poor, we always seemed to have a shilling or two for these special occasions.

On the farm working with the men, I listened to their tales of hardship, how some of them fought in World War One. Now they had to work long hours, from six in the morning until five at night, and were lucky to have a job at all.

In the autumn the fields had to be ploughed and made ready for the next summer crop, poultry and cattle had to be fattened for Christmas. In the winter the farmyard was cleaned out, hedgerows trimmed and ditches cleaned, the cattle were always given a nice warm bed of straw, the grain in the barns dressed, weighed and sent to market. In the spring the summer corn and seeds would have to be sown, then these fields had all to be kept free from weeds, beet, turnips and mangles hoed.

No farmer could be idle, and preparations had to be made months ahead; he could not afford to employ men who were not prepared to work together. Wages were low, which meant that there were two or three other men ready to jump into one's shoes. This was the life I found myself in when I started to work. I enjoyed the life but much troubled me, my head was full of resentment against the rich. It was the rich who owned all the land, farms and houses and who drove their cars to races, but what angered me most and filled my heart with bitterness were those big gentlemen who arrived down from the city in their big cars to shoot our poor little animals and birds. Many a life have I saved as my keen eye picked out a bird or hare trying to hide its terrified self behind a leaf or bramble. How pleased it made me feel to think that I had saved its life and a country boy had deprived the city gentry of some of their sport.

Not surprisingly, I turned towards the Labour party, who held meetings on the village green and promised us better conditions, less hours to work and more pay and always finished up by singing the 'Red Flag'. I joined the union, paying a subscription fortnightly at one of the village pubs. There we would all gather, pay our money, join in a game of darts or cards and discuss the topics of the village news over a pint of beer. Once a year we would hold our union supper. This was a great occasion for me, one that I loved and cherished, though not for its political bearing.

Quickly the time passed. I had gained my place in the cricket team, played football and looked forward to the coming sports; at work, I could take a team of horses to plough or drill the corn. I could plough a straight furrow, and judge the straightness of the corn drills, which went a long way in the estimation of your ability as a farm worker.

Life in the country was slowly changing. I don't think at the time I noticed it. The roar of the tractor as it turned three furrows to a horse's one. The hum of the motors which were filling our roads. More people were taking their Sunday walks to the main road to watch the traffic from weekend trips to the seaside. One or two of the village boys had motor bikes which they rode around the village green to the admiration and envy of most of the others but to the accompaniment of remarks, from the older men 'Those things ought never to be invented'. In our clear blue skies a silver object now flew. Often as we worked in the fields we were attracted by the drone of a plane. We would lean on our forks or hoes, gazing upwards and watch these machines passing over; the older ones would shake their heads and say: 'If we had been meant to fly the Lord would gave given us wings.'

Personally, I was all for progress and often wished I could go back to school, or better still have a private tutor, one of whom I could ask questions without the fear of someone laughing at me as I was still reserved and hung back when in company. It wasn't so much the difference between rich and poor, it was conversing on an equal basis. I wanted to do something, go somewhere to help others, and wondered why we didn't all have an equal chance in this world. I found myself tied to the farm with long hours and hard work, without any prospects of ever becoming a farmer. I wanted to progress.

I was now in my eighteenth year. Life was great. I was a regular member of the cricket team. On my racing cycle I had given a good account of myself, as I could in nearly every sport, and enjoyed doing it. I had never been out of work. The land made you feel you were doing a worthy, helpful and certainly most useful job. Why then was I not content to remain, fulfilling the usefulness of my life? I had been reading books of daring sea stories. Men had risked their lives that this land of ours should remain for ever in our hands. Perhaps it stirred some adventurous spirit within me – I know not what it was, but jumping on my racing bike one Saturday afternoon, racing the bus into town, I sought the Royal Navy recruiting office, and before I knew what had come over me I was gingerly knocking on the door of a house where I had been directed. It was quickly opened and the firm voice said 'Come in, my lad, and take a seat.' 'Now what can I do for you?' said the same firm voice. 'Please, Sir, I want to join the Navy.' 'Right, son,' said he. Now looking round the room I saw two more young fellows were seated on chairs. 'Three of you,' said the man, reaching for some papers. 'I shall want you to do some dictation and arithmetic.' Handing us each a sheet of paper he then proceeded to read a paragraph from a book, followed by four sums. Now I had hardly touched a pen since leaving school. How awkward I felt with a small pen in my hand, and it made me feel more so when a large blob of ink appeared on the paper. After giving us plenty of time for this task, and after we had written our name in capital letters on our respective sheets, the recruiting officer then collected the papers and disappeared through the door. Uneasily I looked at my two companions, whom I thought had the same uncomfortable feeling as myself. We still had not reached speaking terms when back he came with those three sheets, which our whole life seemed to depend on.

Addressing the other two lads he said: 'Not too bad. I will enter you as Stokers.' My heart stopped beating as he turned to me. Did it mean that I was not good enough? 'Now you, your paper is good. You can enter as a seaman.' Puzzled, but waking up to the fact that I had been accepted – what the difference was between a stoker and a seaman I didn't know, but I assumed it must be a bit higher than a stoker – we gave our addresses, signed forms and were told to attend for a medical examination on the following Wednesday afternoon. This created a problem for me as I turned it over in my mind. 'Sir,' said I, speaking up now, encouraged no doubt by the fact that my paper had been the best of the three. 'How shall I get the afternoon off? I don't want my master to know what I've done.' 'Tell him your Granny has died', was the unhelpful reply. Now the farmer knew all my relations as well as I did. He knew that I had no Granny. Hesitatingly I turned away.

5

Better I thought to let this problem sort itself out. The three of us finding ourselves out on the pavement, quickly became friendly, and before parting declared we would all see each other at the medical examination.

This now left me the explanation to my parents. Slowly cycling home, turning it all over in my mind, this I thought was the best argument. 'Dad, the birds leave their nests and have to fend for themselves. At the age of eighteen isn't it time I left home to make my way in the world?' This I considered was my best defence. Arriving home, I at once informed my parents what I had done. How my poor mother looked at me as she remarked, 'You have had a good home and been well looked after. Now you want to go and leave us.' I did not want to leave home to hurt their feelings. In the back of my mind I wanted to prove to them what a worthy young man they had brought up. Somehow, making this clearer in the best way I could, I won them over.

I went about my work busily for the next few days. It remained for me now to get the afternoon off. This worried me, but on the Wednesday morning, finishing the job we had been on for the last few days, the opportunity presented itself. As the boss scratched his head to find us a suitable job for the remainder of the day, I simply asked for the afternoon off, which was granted without any fuss. The only curiosity shown was by his son who said: 'You going to join the Army?' This was emphatically denied by me, although so near the truth. It only remained for me to present myself that afternoon for the medical, and once again the three of us found ourselves together awaiting the verdict of the doctor who soon informed me that I and one other had passed – the other poor chap had failed. We both felt sorry for him, but I am afraid this was soon forgotten in the elation of our success. Now we were informed that all that remained for us to do was to await our call-up, which would arrive by post in due course.

Once again I found myself back on the farm, when one morning the farmer came over to me saying, 'So you are going to leave me to go into the Navy. I've had papers come this morning for your character. Well I don't blame you and hope you do well.' This put my mind at ease, I could now tell everyone, and looked forward to that great day in my life which was to put me on a new road, to be on my own and meet the perils of the world. As I went about my work in those last few weeks I wondered what lay in store for me. My friends and those whom I had played with would no doubt see me from time to time. My childish kingdom where I had dreamed my dreams would never vanish from my mind, and if God spared me I would come back to roam these woods and entwining paths again, to watch the flocks of birds sweeping through the wintry sky. These were my thoughts as that fateful morning arrived when I said goodbye to my father, mother and sisters on the 6th of March 1933.

CHAPTER TWO

The train rattled and shuddered; I sat in the compartment watching as woods, fields, houses and, at intervals, whole villages went flying past. I was now alone, really on my own. Now the villages grew bigger and rows and rows of nothing but houses appeared. The train slackened speed. 'Is this London?' I asked anxiously. 'This is the outskirts. We won't arrive for another fifteen minutes', was the reply. I settled down again and fell to wondering where all the people went to; who lived in these houses, where did the boys and girls play their games? I had not seen so much as a tree for the last ten minutes. Now the train was coming to a halt, and I found myself on the platform with everyone going in one direction. In another minute the rush of the crowd carried me through a small gateway. where a man grabbed my ticket. Bewildered, but still following the crowd, I found myself on a wide pavement with an endless stream of buses and cars going in all directions.

At last I stood in that great City of London with its roar of traffic, its many people going to and fro, and had to find my own way in the busy maze of streets.

My orders were to get a number 11 bus and ask the conductor to put me off at Whitehall. A few minutes wait and here was I jumping on a bus; there followed a series of stops and starts for about twenty minutes. Turning to me the conductor with a nod shouted 'Whitehall'. Jumping off the bus I looked around. There was no mistaking this famous building. I mounted the steps and passed through the doors, presenting myself as having come to join the Navy. Particulars were taken and I was shown into a room with about twenty more young fellows of my own age. Then all through the day followed tests of all kinds – hearing, colour, vision, education, interviews; doubts and fears arose within me. I seemed to be the only country boy amongst the lot and I was self-conscious with all these strange questions being asked me.

It was late in the afternoon when suddenly my name was called out. As I answered my heart sank. Already several had been turned down in a short while, when a cheerful voice answered saying 'You've passed'. What a relief it was to hear these words. About a dozen of us were put in a lorry at the end of the day, rushed across London and put on the train for Chatham. There followed two more days of inspections, tests and interviews, with another two young fellows being sent back, but at the end of it I found myself duly signed to serve my country in the Royal Navy for seven years regular service, and five on the reserve, commencing as Ordinary Seaman SSX 13707.

For the next two weeks we carried out Joining Routine. This took us over the whole of the barracks. Every office had to be visited. We had to be put in a watch for leave, paid, visit the Sick Bay and Dentist, had our civilian clothes packed and sent home as soon as we put on our uniform, were taught how to string our hammocks, as no one slept in a bed, how to do our washing and mending – no running to Mother – we had to do it ourselves, draw our meals at the galley, wash up, scrub and clean the Mess Room – this was all thoroughly organised.

Chatham Barracks was built halfway up a steep hill, consisting of four large double blocks named after famous admirals, the first two being called 'Nelson' after the most famous of all. These were the Officers' blocks, then followed the others, in seniority until the last one, which was called the 'New Entry Block'. We, I expect, straight from civilian life, were unworthy as yet to have a block called after a distinguished admiral. A road ran both sides of the block, the eastern side being the main frontage with its wide terrace lined with trees, a wall and a fence built along the edge of the terrace with three lots of steps to take one the thirty to forty feet down to the parade ground. Across the parade ground was the drill shed, which was almost the same length.

The whole barracks was H.M.S. *Pembroke*, and when one went on leave, or went through the Main Gate, it was always referred to as 'going ashore'. Right at the other end of the barracks, past all the blocks, was the Gunnery School, where no one dared enter, except at the double, and every other movement of one's body was preceded by a loud, sharp order.

All this I had to take in or try to work out in my first fortnight. I did not have time to reflect on what I had done so far. Life was moving fast and I was running to keep up with it. For example, on the day we joined, the officers advised us if we had any money to give it into the safe keeping of the New Entry Office. As I had 30s I decided to put £1 of it away. Meeting a Petty Officer on the step leading to the door and explaining my mission, he kindly took it, saying 'All right, son'. Now the only thing I retained in my mind was his rather large nose. About a week later, when I needed the money, I was rather confused in telling one from the other, but after carefully watching during our work, I selected the one whom I thought had the biggest nose and boldly asked for my £1. After some time in going through his pockets he handed it to me, muttering something about not wanting to lose his badge. I turned away, somehow feeling I was lucky to get it back.

From the windows of the block I had my first view of a naval dockyard with a large number of ships of all sizes. Their masts and yardarms gleamed and sparkled in the bright sun. Some of these ships were painted grey, others a gleaming white, the grey being the Home Fleet and the white going to or from foreign stations. The River Medway flowed steadily by in a half circle past the dockyard. The sound of busy humming and the sound of a siren would occasionally reach us, and our imagination would take over as we decided which ship we would go on, and where we would like to go, maybe China, the West Indies or perhaps the Mediterranean. We were already beginning to feel our sea legs, although as yet we had only seen a ship from the window of the block.

As I finished my Joining Routine, with my name darned in my socks, etc. or stamped on every piece of clothing, having learned to sling my hammock, change and scrub the dirty one, I found myself in a class of twelve, named after Admiral Howe, starting an eight-week course of discipline under the instruction of a Petty Officer.

Every morning after being called, tumbling out of our hammocks, storing them away, breakfasting, washing up, cleaning and polishing the whole room, we were mustered by our class leader – he being one selected by us as leader – and reported to our instructor, who in turn reported to the senior Chief Petty Officer. After eight weeks we thought we were the pride of the New Entries Block. With heads up, chins in, rifles held straight, our arms swinging, in perfect step, we passed our first parade examination as Ordinary Seamen.

The next part of the course was to train us in seamanship. I did not know what I had let myself in for, all I knew about a boat was that it floated on the water. One morning we were marched from the barracks into the dockyard, to be shown over H.M.S. *Marshal Soult*. There we learned about the ship's anchor cable holders, capstan, mast and yard, the rigging, the different kinds of boat which the ship carried, how one hoisted ropes, wires, parts of the ship, foc's'le, midships, quarter deck, bridge, port and starboard sides and so many other things which came under Seamanship – sailing, boat pulling, navigation. My head was in a whirl, I had no time for going ashore, had to knuckle down and get on with it. Besides all this, there were the swimming baths, and five of us could not swim at all. Now when we went for our swimming class I had to jump in the deep end and at the end of a struggle a sailor with a long pole stretched it out for me to grasp. To pass out in swimming, one had to put on a canvas suit, swim three lengths of the bath and float around for three minutes, besides jumping from the third or top diving board. This I felt I must pass, so almost every evening I found myself at the swimming baths, struggling in the water. It took me three months before I could swim a stroke. After that I gradually progressed, and in another two months I felt I was capable of passing this test, which was accomplished without much trouble. I had the satisfaction of being the first of the new five to do this.

Boat sailing was another practical lesson we had to be taught. We were taken on the Medway, taught to tack and go about in a whaler, to rig the boat for sailing with foresail, main and mizzen. Then we were taught boat pulling.

Much more interesting was the instruction given us in seamanship, learning boxing the compass, steering a tug up the river by day and night, the different navigation lights, the buoys marking the channel, the lightships, which side to pass on, how to swing the lead to take a sounding of the depth of water, to handle large wires and ropes and their maintenance. We learned the many flags and pennants of the Navy and other countries. We learned how to send messages by semaphore and to read them, the morse code, and, of course, knots and splices. When we had gone through all these subjects and some more which I can't remember, and swotted night after night, going without our night's leave, even sleeping on our Seamanship Manual, the week of passing out on each subject duly arrived. One by one, we were called in to the Office and given our results. With great relief and satisfaction, I learned that I had made it, but with very little to spare. One of our classmates had to revert back to another class.

Now I really felt I was in the Navy. We still had a lot of training to do, but the major examinations had been passed, and the fear of being sent home as a failure had gone. Stepping out like an old salt, although we had not yet seen salt water, we felt we would like a run ashore to see something of the town. Wouldn't the girls fall for us in our uniforms? With special care we tilted our caps to the right angle with the bow right to the front, and prepared to take Chatham by storm. We filed through the main gate on to the dockyard bus. We strolled through the town, calling in at the Navy tailors who knew how to cut our suits as we wanted them (not as the standard laid down), then to the photographers for a photo to be sent home, then a pint of beer at one of the pubs. Here the barmaid seemed to know more about the Navy than we did. She referred to us as 'Sprogs' (a term applicable to one who has just joined), and said 'Have you been home to your Mummy yet?' When a couple of sailors with badges on their arms, and that weatherbeaten complexion that only men who face the hazards of the sea and are tanned with tropical sun get came in, her interest in us died, so we drained our glasses, and went to a show, finishing up with hot dogs at the 'Sailors' Rest', before walking back to the barracks and turning into our hammocks. We had been taught that we had a long way to go before being acknowledged as Sailors of His Majesty's Navy, although we wore the uniform.

A few light days of work around the barracks, then we had to forget all about seamanship. 'Put it at the back of your mind for a future date', was the advice of our instructor, 'as now I am going to turn you over to the Gunnery School to be shaped into a smart body of men. So far, you have been spoonfed, and the Navy is not a nursery, so from now on you come under the Gunnery School. This is the first and last warning.' The next second he let out a bellowing yell, and we all sprang to attention with such amazing alertness that it would have done justice to the Marines. The order was obeyed with such alacrity that number three's hat jumped off his head, with the sudden shock. When asked to account for it, he made the remark that he thought the Petty Officer had stabbed himself with his bayonet. This only resulted in the whole of the class being marched up and down at the double for ten minutes.

After a few days we were issued with gaiters which had to be blancoed and brass tips polished, with belts and rifles to draw each morning and fall in with the gunnery platoon on the parade ground. The air of the officers, and everyone connected with the Gunnery School, seemed to have that snap about it, that upright bearing of importance which comes to those who just give orders, especially knowing that no one dare disobey.

Whenever we came into the precincts of the Gunnery School it was always at the double march, entering or leaving, and one became more like a number. It was full of almost every type of naval big guns; we had to be taught everything about them, and each man's job at each position. These all went by numbers, and all one heard in the morning was 'Nos. 1, 2, 3, 4', snapped out in rapid succession, 'Right. Gun ready. Fire. Change Rounds. 12345 Ready' as we moved at the rush from one gun position to another.

There were many things to learn about gunnery. We had to know something about each, from a revolver to a 15" gun which fired a shell weighing a ton. Thinking I was a crack shot and given a loaded revolver to hit the figure of a life-sized man at twelve paces, I missed every shot. We learned about the rifle and its many uses in drilling and on the range. The rifle range was at Sheerness, which meant a trip up and down the river each day in a launch, which for most of us was the longest boat trip we had ever taken. We saw the inside of a 15" gun turret, the biggest gun in the Navy, with its amazing mechanism, the elevation and tracking dial which we had to learn to read, the loading by power of its shell and charges. The many other guns we had to know something about kept me so busy and occupied in trying to absorb this new knowledge that I became accustomed to my transformation from the comfort and security of my inland country village to the busy, thriving centre of one of Britain's powerful seaports.

As we stood to attention on the parade at the end of our extensive term of training, our backs straight, heads held high, we felt worthy members of this Senior Service, and a credit to our country. Our instructor stood us at ease, and the Captain of the Gunnery School made a short speech complimenting us on our ability to adjust ourselves to the stern discipline of the last few weeks, and told us that we would be a credit to our friends and family, and always to carry that upright bearing, the mark of the Gunnery School. With firm steps and arms swinging we marched off to hand in our rifles, belts and gaiters to rejoin the barracks.

CHAPTER THREE

For the next few days we were to proceed on our summer leave. Collecting our money, railway vouchers and leave passes we busied ourselves with our best Number One suits, carefully pulling our collars into place and adjusting our silks. The road was thronged with navy blue, the trains in turn being loaded down as sailors hung from the windows of the carriages. Not until Victoria Station did the crowd absorb the boys in blue, but even in London's many stations there were always two or three to be seen.

Now the train puffed to a halt, and I stepped on to the platform. As I had now travelled I was more of a man, and wanted to show off my manly bearing. I strolled nonchalantly to the bus which would pass by our country cottage and took my seat next to the window. The bus slowed up and came to a standstill. I jumped from the bus to pass through the little gate, round the house, to enter the door where all the family were so eagerly awaiting my homecoming.

A trip was taken down to the village with everyone asking 'Now, do you like the life?', with a drink and a game of darts in the pub. A dance at one of the village halls ended in a moonlight cycle ride in the early hours of the morning, with a lie-in until noon. A visit to town with young ladies running to touch your collar for luck, watching a film or going to the theatre, tea at one of the small restaurants. Very quickly my fortnight's leave slipped by and I found myself saying 'Goodbye'.

'You are back in the Barracks and down to work. Forget about your romances and let us have some signs of life. Pay attention or we will see what a little double marching will do.' This warning quickly pulled us up, and our eager chatter came suddenly to an end as we formed up in our class to carry on with our training, our leave finished. There was still a lot more to do to complete our course. The next item on the list was the torpedo class. This dealt with the actual torpedo as well as with the whole electrical fitting of the ship – the guns also being electrically fired – and of course with all the communications, and with the telephone system with connections in all compartments. This course was more relaxing, and an understanding of the various instruments, their names and where they were used was enough to qualify.

As we progressed through our course and became more accustomed to the Barracks routine, we learned where to dodge to get out of certain musters, how to skip the church parade with more time to go ashore. I well remember one Sunday afternoon when three of us explored the Medway towns, and found ourselves at Gillingham Strand on the bank of the Medway, with its swimming pool, putting green, small railway and boating pool for children, a bandstand, ice cream, snacks and small beach, with a view of the dockyard with its masts of many ships. The charm and peacefulness that the ordinary family looks for.

The three of us mingled in the crowd, dodging here and there, eating our ices, as three young girls, also out for the afternoon, somehow mixed with us as we smiled, played and chased around in the sun of the beautiful afternoon. One of them attracted me with her lovely curls, hanging in loose braids over her shoulders. As she laughed and ran they gathered and tangled as she twirled and turned amongst the people, her sparkling eyes shining through the fringes of hair that the light breeze rustled across her brow, held a challenge to run and catch her as she quickly eluded me in my rather bashful approach. A dimpled smile danced upon her face as she brushed her curls away. She laughingly spoke to me. I cannot remember the words she uttered, in her soft, gentle tone, so English, yet in her merriment giving away the fact that her home was overseas. Soon we tired of running about and we lingered on the sands that had been sprinkled over the pebbled beach. We played noughts and crosses and asked each other questions. The three of them were sisters and had not been in England long. Their father and mother came from Kent and had emigrated to Canada some twenty years before, had met and married out there, and now had brought their family back to England. We enjoyed a pleasant afternoon, and waved farewell to the young ladies as they sped off after their tea in a light-hearted way. We wandered back to the Barracks in our own good time, talking of our next week ahead when we were due to start a week in the galley, with the cooks teaching us victualling arrangements on board ship, how to prepare food and cook it, the general messing system on a ship, and the canteen messing on the smaller ones.

One unfortunate incident happened to me at that time. One morning, feeling unwell, I reported to sick bay. I was put to bed and then rushed to the Naval Hospital and put in the ward. There I discovered I had bronchitis, with a high temperature. After a week or so in bed I gradually improved, and after a day or two up I found myself becoming a Nurse Orderly, waiting on the other patients, cleaning the ward, serving the meals and washing up. I never saw a nurse, except when the Medical Officer came on his rounds once a week, and I don't think I got a discharge until there was someone else well enough to take my place. This put me back, so when I rejoined my classmates they had all been drafted to destroyers which were being commissioned for the Mediterranean. I joined another class who had finished their training and were waiting to join a ship.

Daily we paid visits to the Drill Shed where lists of all drafts were put up. One morning, some of the lads came rushing up the stairs to inform us we were all down to join H.M.S. *York*, a cruiser going to the West Indies station. She was refitting in the dockyard, and we would be an advance party. Now we were all greatly excited, and that night we sat down to write home to tell our friends and relations we would be joining a ship in the next few days. Some also rushed ashore at the first opportunity to buy cap ribbons to have them altered to slip on when they stepped aboard. Each day we watched the board for further news.

It was in one of these dramatic moments that the loudspeaker buzzed, calling 'Attention'. Now everyone listened as the message came through: 'Will all ratings on draft to H.M.S. *York* muster in the Drill Shed at 2 p.m. to draw their tropical clothing.' We let out a cheer.

Soon everything was completed. Our clothing with wide-brimmed helmets was packed in our kitbags, we had visited the dentist and sick bay and been inoculated, vaccinated, and had our cards stamped. Now all twelve were ready to march into the dockyard to join our ship and become real sailors. We marched to the farthest dry dock, and there lay the *York* all rusty and dirty, with dockyard workers swarming over her. The only thing of note to be seen was its very tall mast. Our class leader reported to the

Quarter Master 'Twelve Ordinary Seamen joining Ship'. 'March them aboard and report to the Master at Arms' office.' This was the last time we were together as a class, and as each of us in turn reported we were given a card with the part of ship, i.e. midship or quarter deck, number of mess, starboard or port watch, etc. Now I found myself in no. 12 mess (starboard watch, 1st port) (port of ship) (midship), allocated 157 locker to stow my kit and hammock, reporting to the leading hand of the mess.

Talking to the Able Seamen who had travelled the world, I learned of the hospitality given to them by the Americans and our large colonies – the offers which had been made to them by some big rancher to desert their ship and marry their pretty daughters. I heard of the golden-haired girls of the Scandinavian countries with their blue eyes which matched the clear blue waters of the beautiful fjords. All these marvellous tales, told with good humour, which are only collected by a seafarer with years of travelling, never failed to fire and hold the imagination of us young salts.

In this, my first ship, life was turning out far better than I had ever imagined. I had not gone into the Navy expecting to find life easy, indeed I do not think I had any expectations at all. If there were, it was to see the world, and I had no idea what there was to do, nor had given any thought to that side of it.

The ship would be refitting for two or three months, then would have to do trials and working-up routine before sailing on commission for the West Indies station. This gave me the opportunity to understand something of my new life, as most of the ship's company had still to be drafted to the *York*. As we were yet only a skeleton crew, allowance was made in discipline and for the condition the ship was in. I had hardly learned which part was which on the upper deck when a Chief Petty Officer came along. 'I want you watch-keeping down in the Telephone Exchange,' he said. 'I have never telephoned in my life yet, Chief,' said I. 'Come along with me, and you soon will,' was his reply. He led me down two decks, through hatches and down ladders, to what I thought must be the bottom of the ship. As he pointed to yet another ladder which led straight down to a small cabinet, he said 'Report to the Leading Seaman, he will put you right,' and he turned away. Scrambling down the ladder I reported 'New Telephone Operator'. I then entered another little cabinet, an offshoot from the other. There sat a man on a swivel chair, with earphones on and a panel in front of him with a large number of leads which he either plugged in or took out as various lights came on or went out. 'You will do both Dog Watches,' remarked the leading hand. 'You're in three watches and close down for the night at pipe down.' 'But I don't know anything about telephones,' I remarked. 'You can spend the afternoon watch down here with your opposite number and he will teach you.' With those orders I thought I had better make the best of it, and learn what I could as I watched my companion shifting plugs all over the panel. After a while he said 'Take my place', indicating his chair, and placing the headphones over my head. A few minutes later a light flickered. 'Put a red plug in [a receiver] and say "Exchange".' 'Bridge,' said a voice. I repeated this. 'Take this black plug [a transmitter] now, plug in, and turn this knob. When you see both lights on they are connected and talking to each other.' After a little while I became quite good at it. The shore telephone was more complicated, but after being shown, and being interested in the job, I became proficient and to my delight I found myself connecting calls from the ship right up to Scotland.

This job I held until the ship became almost ready for sea, and the ship's company was completed and I had to take my place to learn seamanship the real way and practise my gunnery. I discovered there was far more to gunnery than actually being a member of a gun's crew. The target had to be sighted, the rangefinder had to give the range and the control position had to add up the range bearing, wind, temperature and lots of other things which went into a big clock placed down in the lower part of the ship, which then transmitted to dials on the guns an elevation and training.

Each part of the ship had its own mess deck and the whole of the ship's company from the Captain downwards had their own parts of ships, mess decks and compartments according to their rank and rating. The officers had the after end – the blunt end – and were entitled to cabins and the whole of the quarter deck to exercise themselves. If the Captain came up from his large, airy cabin, he alone had the starboard side of the quarter deck. The ratings all lived for'ard in their respective messes. The chiefs, who were the senior ratings, were on the lower deck, followed by the Petty Officers, who had 'B' gun deck to pace up and down and enjoy their leisure time.

The rest of the ship's company, consisting of marines, stokers, sick berth attendants and all the seamen, boys and Ordinary

Seamen like myself, who were the lowest of the low, were given enough room to eat on the foremost mess deck where the ship pitched and plunged its nose in the heavy swell, and at sea all portholes had to be closed. For sleeping our hammocks had to be slung above and below each other and the tables and forms were made use of as well. The snores and noises that issued from the throats and nostrils from so many sleeping companions had to be endured. The odour from the socks and feet of the hardened matelots made us sleep all the more soundly.

When the ship's company was completed and the dockyard workers had finished the refitting, I found I was learning the meaning of words of a different language. For instance, the Master at Arms was the 'Jaunty', the Bo'sun the 'Buffer', the Leading Seaman 'Killick' and so on throughout the personnel and everything connected with the Navy. A new coat of paint was given to the ship which changed her altogether from a dirty, rusty hulk into a ship I was proud to belong to. The sailors wore their suits instead of overalls, the marines took their place on the quarter deck. There was a smart salute as one stepped on the ship and another as one stepped on the quarter deck. When I asked 'Why?' I was told it was in memory of Nelson. There was a general air of smartness and the Marine Band and Guard were present at colours each morning.

Rumours grew fast all over the ship that we should be leaving harbour every day for our trials. Most of these rumours started from the galley; Cookie had overheard the Master at Arms say there would be no leave tomorrow night, or a signal had been sent cancelling a dinner engagement so some enterprising Able Seaman had put two and two together. Now we would be going to sea. These rumours persisted throughout my whole naval career. Sometimes they sprang from wishful thinking, or a pessimist might be foretelling our future movements, but later on I came to take it as part of the life of the Navy. Eventually a notice was put up informing everyone that the dock would be flooded at 10.00 hours. We would take berth in the basin, and in the next couple of days would be proceeding to sea for trials and working-up routine. The dock was flooded and we entered the basin towed by a small tug and took up our berth, secured by wires to bollards on the jetty. Our job was to man fenders, large round mats of coconut hair which we lowered over the side, or man the wires, taking up the slack or easing out, to whatever order given.

A large number of ships were laid up in Chatham Dockyard, from small minesweepers to large cruisers with a number of submarines, and on the boat deck of the *York* we were working, four young seamen with a young Petty Officer in charge. It was a nice sunny afternoon. Soon we were all leaning on the guard rails watching a submarine diving, the periscope rising out of the water, then lowered, the sub occasionally breaking the surface. In the morning, however, the P.O. was sent for by the Commander, much to his surprise, and given a thorough dressing down for not keeping his men at work and actually encouraging them to be idle. The brave P.O. denied all this, and stuck up for himself and the men, and wondered how the Commander had learned all this and what sort of shipmate had betrayed him. Shattering the P.O.'s excuses into a final silence as though a torpedo had struck him in the engine room, the Commander retorted 'Rubbish. I was watching you myself from the periscope of the submarine all the afternoon.'

Everything was now ready for sea. The band of the Royal Marines were assembled in the quarter deck. From for'ard to aft the seamen were lined up. The Captain with all his officers in attendance was in command of his ship, from the bridge. Flags fluttered from the yardarm. Bugles sounded and the shrill sound of the Bo'sun's call echoed over the waters as each ship saluted the other in order of seniority of their respective captains. The officers' wives with friends assembled on the quay, waving farewell. 'All lined up in seniority from the Captain's wife downwards,' whispered a three-badge A.B. standing next to me. This I doubted, as I had recognised one of the ladies from one of the bars frequented by matelots.

The *York* was now at sea under her own steam, and this was my first ship and sea-going trip. I watched the Barracks and the dockyard fade away. We swept past the strand which made me think of the young girl with the fair curls and mischievous smile. Quickly we stowed away the fenders and reeled up the wires, putting their sea-going covers on. The band and all other ratings had been dismissed. Our sea duties amounted to lookouts close by the bridge, reporting all objects that came into view. The Bo'sun's Mate broadcast the ship's routine and all special orders from the loudspeaker system, or manned the telegraphs from the wheelhouse. These telegraphed the speed of the engine as required from the bridge to the engine room. These duties we carried out in three or four watches throughout the day and night. In the forenoon there was the mess deck and part of the ship to keep clean, and throughout the day and night we practised at our gunnery position.

CHAPTER FOUR

My first experience at sea was not all that encouraging. The sky was grey and the sea seemed to be more grey than the sky. The wind was cold; it slashed the waves and made them wild and angry. The further we went out to sea the more the ship seemed to pitch, and quite a number of us gathered amidships, mostly very quiet. I wasn't feeling well, and when I looked at the others, some with years of experience, they all seemed to have that same feeling judging from their white faces. I didn't go for my supper, neither did I turn into my hammock that night. It was the first time of many that I wished I had never joined the Royal Navy.

Next day we anchored off Portsmouth. The booms were swung out and the motor boats lowered for the Captain to pay an official visit ashore. Now we had settled in more sheltered waters I began to feel hungry and found my way to my Mess, to be asked where I had been for the last 24 hours. What was I going to do when we really did run into some rough weather? What about the time we ran into a typhoon in the Pacific and all the boats were smashed? I let these able seamen run on, taken up with their own stories as I made myself a nice big sandwich. Sea-sickness had its compensations; you could appreciate a good meal when we reached harbour.

We weighed anchor the following day and went out to sea for speed trials on our way back to Chatham. This mostly concerned the Engineering Branch. To the discomfort of all, the whole ship shook and shuddered as we gathered speed, cutting through the water with spray drenching the bridge, the stern settling low down in the water to throw and churn a wide wake. The mast and rigging seemed to strain and bend under the stresses put upon them, with the funnels belching black smoke which hung low over the water, fanning out in a wide arc as if we were leaving a black thunderstorm behind us. The little ships and boats rocked and swayed in the waves that swept from either side as she sped on her way.

At last the ship slackened speed. The sea was now much calmer than when we had started off four or five days before. We were approaching the mouth of the Medway on our return to Chatham. Special sea duty men were called to close up at their stations (this is always done when entering or leaving harbour). I was down in the Mess clearing up the breakages after our adventurous sea trials. Soon everything was shipshape again. 'Stand easy' sounded over the loudspeaker, and everyone gathered in their mess for a cup of tea. We were all looking forward to a run ashore at Chatham, and had so much to say to each other that we failed to hear the loudspeaker order to carry on working, and not until a P.O. with a notebook and pencil poised had already entered our mess did we make a hasty retreat.

We made our way on to the upper deck to get all the wires and fenders ready to tie up alongside the jetty, and then we lined up again from for'ard to aft to enter harbour, the bugle sounding as the Captain called us to 'Attention', then 'At Ease' as the 'Carry On' sounded. The saluting over, a tug gently pushed our stern in as we threw our lines, paying out the wires and ropes for the dockyard workers to make us secure and run the gangway inboard. Soon the Quarter Master and Marine Guard had taken up their position with the young bugler in attendance. A few stores arrived which I helped to carry on board before our day's work came to an end and the loudspeaker finally said 'Hands to tea'; 'Liberty men will fall in in fifteen minutes' time'. That ended my first sea-going trip. I now felt a bit more sure of myself, as if I were beginning to get my sealegs. I could write home and tell them all I had been to sea for almost a week out of sight of land with a big sea running. No, I didn't think it would be necessary to tell them about my seasickness. I thought I wouldn't mention it.

Very soon now we would all be ready to proceed on our long commission 'to the far-flung corners of the British Empire' as one of the young A.B.s put it. Everything required for these warm climates had been brought on board, including ammunition, for at that time many a difficult situation had been overcome by the mere presence of a ship as it dropped anchor in some remote bay.

Nearly everyone was going to have a final run ashore. I seemed to fit in well with my shipmates, although most of them had done as many years at sea as I had done months in the Navy; I felt rather honoured when they turned to me saying 'You can come too if you want to.' My number one suit was taken from the locker, pressed and ironed with as many concertina presses in my wide bell-bottomed trouser legs as I could get, my cap ribbon with its dainty bow now read 'H.M.S. *York*', which I thought would impress the barmaids in Chatham. Already I was thinking I had the rolling gait of a seagoing matelot.

We picked Saturday night for our run ashore, and made arrangements to meet at one of the well-known pubs for the final farewell. We caught the Liberty boat at 17.00 hours. We mustered by the gangway to be inspected by the Officer of the watch, the Leading Seaman facing us other ranks. The inspection was soon over and the P.O. reminded us that leave expired at 07.00 tomorrow morning. Then we filed down the gangway, picking our way through the dockyard and out of the gate, under the ever-watchful eyes of the Customs Officers, who seldom ever worried us as we usually kept to our quota of cigarettes. We boarded the bus and were away into town, to make our first stop at a restaurant, where we took up half a dozen tables.

Our main entertainment was to be at an old theatre. One of the turns at this show was the generous offer of £5 to anyone who could ride on the back of a mule for three minutes. On our way from the restaurant to the theatre we called in at several bars, and we were all merry and excited with the thought of what was going to happen as we took our seats, filling two rows. The show was good with plenty of good jokes – mostly about sailors as I think the comedians knew their job and what was expected of them in a naval town. Now came the turn of the mule with the challenger making his speech, offering five £1 notes as he spread them in a circle, dazzling his audience with all that money. Up jumped Jolly Jack, willing to have a go at anything, to the cheering of the crowd and to the extra cheering of his shipmates, until 10 or 12 stood on the stage ready to have a go, and the old mule stood there as though glued to the floor. Our main hopes rested on the last man, as he knew something about mules and donkeys because his grandfather kept them on Yarmouth beach, entertaining the holidaymakers, and before he joined up he used to help him at weekends. We had stuffed his jumper with the largest carrots we could find and bought two pounds of lump sugar to soothe the savage beast. He had told us his plan as we bought him a few pints to keep his courage going, and we were all to share in the £5. He was to be the last one, as the mule would be feeling tired by them, and a large carrot with a lump of sugar at the right time couldn't fail.

One after the other mounted this mule. It seemed such a calm and docile animal until someone mounted its back, then it bucked and jumped and kicked with both legs, twisting and turning, which seemed impossible in such a small space. Soon only one remained, to the beaming approval of the ringmaster, who bowed low with a flourish of his top hat as each bit the dust. Now, with a different approach, our last man took from his protruding jumper such a large carrot that it made the poor old mule's eyes blink as he dangled it in front of him. Now he offered it to the mule to eat, which he accepted gratefully. Jack followed up with two lumps of sugar. Consternation spread across the face of the ringmaster as he began to realise what was behind it all, and he stepped forward to protest. Now it was our turn to protect our hero, which we did with loud approval and declared all was fair, so the show went on to a ringing cheer from the now excited audience.

With lumps of sugar our shipmate was getting on well, now he stroked its neck and patted its back, reaching forward for it to take the sugar. With a light leap he was on its back. All was quiet and tense as the mule turned steadily round looking for its sugar. Breathlessly we watched the clock. A minute had gone, two minutes came up – a vision of the £5 danced before our eyes, when suddenly from the stage door burst the ringmaster flaunting a large whip. Now he cracked it. The mule instantly stood on its hind legs with Jack hanging on for dear life, another crack, down came the mule to roll on the stage with poor Jack completely beaten. The ringmaster, now his old self, made a sweeping acknowledgement to his audience who in their excitement had left their seats, but now, more subdued, were settling down and taking it all in good fun.

We left the theatre rather pleased with ourselves as our plan had almost succeeded. To finish the evening we called at one of the largest pubs. The beer made me feel warm and talkative. I noticed too that a number of the opposite sex had somehow mixed in with us, pushing their small glasses alongside our big ones. Who cared anyway? We were young. Life was gay. Soon we would be far away.

We poor sailors were no match for these ladies of the sea, who knew just when our glasses wanted re-filling, and to save themselves two journeys always filled their own at our expense. Soon the floor of the pub seemed to be rising and falling like the rippling waves at sea. I staggered across the room, fighting my way to the open door. Somehow I reached the pavement, to rest awhile and let the strength get back in my legs. With short bursts of speed, and five-minute rests, I was trying to make my way to the ship, but every now and again the road would rise as if a seawall had been built in front of me. I touched what I thought was the wooden deck and sank back exhausted. A voice was calling: 'Come on Jack, you can't sleep here, it's getting light.' Somehow I raised myself and shook my dazed head. I was lying on a park-bench. A passer-by was helping me. I thanked him as I pulled myself together and decided to walk back to the ship. That would help me to recover. I would have time to wash and clean myself before the ship's company came to life.

The day grew near for us to leave. Our last letter was posted. Only leave up to 22.00 hours was granted the night before. The Master at Arms reported no absenteeism. Early in the forenoon the *York* left in all her glory, her coat of fresh white paint gleaming even whiter against the greyish water of the Medway and background of greenish hills. On board we stood to attention as we paid our last respects to the naval base and all those who waved us farewell. Would they be waiting when we returned, or would they succumb to the snares and temptations that beset their path? Their allowance was so very small in those days, and if weak, they were an easy prey to the prowling sharks of the seaport town. Jack also had his temptations abroad as much so, if not more. He was tempted to enter the smooth waters that appeared to be so calm on the surface, but underneath the tide was so rapid that one could soon become engulfed in a sea of misery. All this I was to learn in due course. As yet I was young and keen, and looking forward to my sea voyage. Our time was taken up by our sea duties, gunnery exercises and keeping the ship clean. In the evening there were games of all kinds being played on the mess decks. I was standing up to the sea fairly well and was only ill when the sea was extra rough.

Gibraltar was our first port of call. The Rock came into view in a blue haze, as we approached it early one morning. We were a long way off when I first caught a glimpse of this majestic fortress of the British Empire, guarding the entrance to the blue Mediterranean. As we drew nearer I could see Africa on the other side of the wide entrance, my first sight of a foreign country. The ship seemed to become so small as we sailed almost under the Rock. Most of the lads had been there many times, and our visit was to last ten days. This gave me sufficient time to accustom myself to the sea and get used to keeping watch at night. During the watch the seaboat crew would be mustered at the beginning and ending of each four hours duty.

Now the ship would soon be secured to the breakwater. Already I had noticed some of my shipmates had made themselves scarce, and only a few of us remained on the upper deck to tidy up and put away all the tackle which had been used to secure the ship to the jetty. Although I was impressed by my first sight of Spain, and across the Straits of Gibraltar the dark mountains of Africa, rugged and desolate, I thought I could see no sign of life in that direction. What mysteries did they hold? Were there really black African tribes with their witch doctors making human sacrifices on the night of the full moon? I stopped and shuddered as I allowed my thoughts to be carried away.

We liberty men reached the boundary of the dockyard, passed the Marine sentry, and crossed the road to enter the town a short distance along the street. We found ourselves in a crowded street of dark-skinned people, gabbling away in their Spanish tongue with here and there a uniformed member of H.M. Forces. The shops were full of curios, wonderful carpets, silks and satins, Chinese and Japanese dinner and tea sets with trinkets of every kind to catch the eye of tourists. We walked the length of the street and turned back – not for us to buy anything at the beginning of our cruise, only perhaps a card to send home. Now the lights flashed on all over the town. The strains of music came floating into the street from many of the brightly lit cabarets.

We drifted in and out of these cabarets, chattering and drinking, joining with all members of the lower deck of H.M.S. *York*, but soon it would be time to end my first run ashore. Already the crowd was thinning, and a large number of patrols now paraded the street, not as I was given to believe to carry back to the ship the young sailors who had fallen for the dark-eyed señoritas, but in most cases for the older two- and three-badge sailors who had fallen for the dancing girls. There were always one or two that no amount of persuasion from their chums could move from their seat, until the patrol led them to the door.

Our stay at Gibraltar lasted four days. That one run ashore had bitten deep into my pocket. Our pay was not generous, and nearly every sailor had some kind of allotment deducted from his pay.

As we young ones had had a taste of our first run ashore, and enjoyed the excitement and company of mixing with the older hands, I thought we had become more friendly. We settled ourselves for the next long sea voyage on our way to South America, during which we could save our money for the next big port of call.

So, saying goodbye to Gibraltar as we steamed past that most famous rock on our way to our next stopping place, we sent our last letters, cards and curios by the closing mail which was collected just before the gangway was hoisted inboard.

I stood leaning on the guard rail, watching Gibraltar getting smaller as our distance increased. There is something remarkable about Gibraltar, some feeling, as if it is the last home port of leave and the first to welcome you on your return. Although it is a thousand miles from Britain it seems to be so safe and secure, and while it remains I think we need have no fear.

CHAPTER FIVE

The *York* had settled down to a steady eighteen knots. The sea was calm and the air cool. A few merchantmen were steadily plodding along, their funnels belching out black smoke, as if they were well laden. Soon we would be out of the sea lanes, well out in the ocean where it was most unlikely we would see another ship for days. We carried out our usual routine during the day, keeping the ship clean, at night watch-keeping.

During the evening and on certain days, tombola, now commonly called Bingo, would be played. Competitions took place between different parts of the ship. A great favourite with the sailors was 'Uckers' or 'Ludo' to you at home, with the boys dressed up in the old clothes of sailing days, blacking their faces and wearing pigtails, having a large wooden bucket to shake the dice, which was almost a foot square. On the messdeck you would come across a small card school, playing their game of solo, and here and there, deep in thought, pen scribbling away, still keeping up their promise to write daily, would be the ones who were so much in love.

On our way to Rio we called at the mountainous Cape Verde Islands, and we were allowed ashore for recreation. It was here I was offered a game of cricket, which I readily accepted. There was an excellent bay for the ship to anchor in, but in going ashore in the ship's motor boat and landing at a creaking wooden pier with a number of natives following our every movement, we were escorted by an official-looking native policeman to a waiting lorry and driven a few miles to their cricket field – a level piece of sun-baked sand and gravel with cement wicket. A laughing eager native team were all ready to take the field; and the next few hours I spent retrieving balls as these laughing natives thumped our bowlers for sixes and fours to the jubilation of the young black spectators. When this polished team did give us a chance to bat, it only took a woolly-headed boy three balls to spreadeagle my wicket, and a big broad grin spread over his face, showing a white row of teeth as he chuckled, so happy and pleased at his success. I limped back to their bamboo pavilion mopping my brow, a puzzled frown on my face, for being beaten so easily at one's own game by a team of foreigners was a bitter pill to swallow. We clambered back into the lorry rather dispirited. I did want to make a show to get in the ship's team at the one game I thought I was good at. Those boys followed us every inch of the way, chattering away in their native tongue. They lined the wooden jetty, their cheering and excited talk carrying across the water as the motor boat sped us back to the ship, a rather silent and defeated team.

One day the Chief gave me a message to convey to the Petty Officers of the different parts of the ship to meet the Commander in his cabin at 15.00 hours. This message I delivered and mentioned to my messmates round the dinner table. The information seemed to arouse their curiosity and I realised that I was, at the moment, the centre of their enquiries. This resulted in a lot of speculation as to what was going to happen; the ship was returning to England, had war broken out or the ship's cat had kittens? Having had our tot of rum and eaten our dinner, we all lay down to sleep, which is the usual thing for almost every sailor, still wondering why the Commander wanted to hold a conference.

With the sound of the bugle call arousing us to muster for our afternoon's work, we dreamily pulled ourselves together as we grabbed our hats to make a rush so as not to be reported adrift. Dismissed from our muster to carry on with our jobs, we were still puzzling over the conference, when coming down from the bridge after doing a spell of lookout Able Seaman Whale had overheard the Navigating Officer tell the Commander that we should be crossing the Equator at about midday tomorrow, and he had also heard them talking of the crossing-the-line ceremony. Now we knew what was being planned that afternoon, or the more experienced ones did, as they informed us we would be shaved and dipped head first into the sea. I was a bit concerned over this as we were supposed to be in shark-infested waters.

We had watched 'Tanky' the butcher bait a large hook with a huge joint of meat, which was towed from the stern for most of the day. We had all watched excitedly when the ship slowed down for it to be hauled inboard while the butcher, who was a Marine (a rather hefty one), stood by with a large axe ready to strike. About twenty of his comrades with short runs (as the space of the ship would allow) swiftly hoisted it inboard. The Chief Buffer had also a six-inch hawser with a running bowline ready to lassoo it if the Marine missed with his battleaxe, and behind these was the Gunner's Mate with a rifle and ammunition, with the best marksman in the ship. All the remainder of the ship's company stood well back, in as much fear of the swing of the battleaxe, lassoo and rifle fire as they were of the shark. After a few sharp runs a lookout who was to give them the down when the shark or bait came into view suddenly yelled 'Shark'. The Marine on hearing the yell broke into an extra burst of speed. The 'Tanky', stripped to the waist with axe held high, his rippling muscles shown to great advantage, arched himself for a mighty swipe, then with a rush the hook came charging over the stern minus the joint of meat, sweeping the legs from under our noble Marine warrior, with the axe descending, its sharp blade going clean through the Chief Buffer's six-inch hawser. A mighty cheer echoed over the whole ship that brought the chefs running from the galley, the stokers from the engine and boiler room. The Marine picked himself up and ran his thumb over the blade of the razor-sharp axe. The poor Chief, rather bewildered, picked up the two ends muttering something which was inaudible to respectable sailors, then he threw them down and marched off as more men clamoured on the upper deck to know what all the excitement was about. That was the first and last attempt to catch a shark I ever saw in the Royal Navy.

We carried on during the afternoon, assembling on the mess deck for our 'Stand Easy' afternoon cup of tea, when we were warned by loudspeaker that all members of the ship's company who had not crossed the Equator before were to give their names in to the Master at Arms Office by 18.00 hours. All ratings with experience of crossing the line who wished to help were asked to muster outside the Commander's cabin at 18.30. Apart from the hands required for sea duty there would be no further work for the remainder of the day. A low ripple of excited comment greeted this announcement as we settled back on our mess forms, or passed the time in the recreation room having a cool iced drink, waiting for the tropical night to descend upon us, when we were to ask permission from King Neptune to enter his domain.

The evening passed, and at last the darkness which so quickly fell after the setting of the sun encircled us. We all streamed out on to the Foc's'le Deck, or whatever vantage point we could find on the for'ard part of the *York*. The ship hove to. Then from the depth of the cable holder a deep gruff voice hailed us: 'What Ship?' Then from the bridge we heard the Captain's answer through his megaphone in his own gruff voice, but not so deep as that which hailed him. 'His Britannic Majesty's Ship *York* does hereby ask permission to enter your domain.' Then from over the foc's'le two stately figures climbed. One, we could make out from the light of the stars, wore a four-pronged crown of silver which sparkled and flashed as he rose to his full height, and covering his breast and back appeared an armour of sea scales which sparkled with the phosphorescence that is ever present in the deep sea. Hanging all round him were great seashells, partly hidden by his long grey beard which reached right down to the deck. In his left hand he held his sceptre, and by his side appeared his Queen with a headdress of silver fish skin which hung down to her waist, covering her bosom and arms; and tapering down to her mermaid's tail were flashing scales of luminous colours with seaweed hanging all over her slender figure. No wonder we all stood in awe and amazement. I really thought I was in the presence of King Neptune himself and his wonderful Queen. As the royal consent was given on the understanding that all those who had not crossed before should be initiated to become loyal subjects; this concluded the first part of the ceremony and King Neptune with his Queen, in stately order, disappeared over the ship's bows and we heard a light splash. Someone said 'it is the tail of Queen Neptune slashing the water'. The ship, after a few minutes, came to life again.

All the ship's company made toward their hammocks, apart from those who were required for duty that night. As I slept in my hammock to the gentle roll of the ship as it ploughed its way through the blue tropical sea, I had a dream of a beautiful mermaid that glowed red, with beautiful smiling eyes which gave me a tender welcome as she sat on a pebbled beach, so that I reached out to touch her reddish hair. Then she slipped away, and only a widening circle of ripples remained. I woke up as the smiling eyes became those of the young girl I had seen on Gillingham Strand and a voice was yelling in my ear from the next hammock: 'Stop pulling my hair!'.

As no more sleep seemed possible and I could see by the clock there was not much longer to go before all hands would be called, I made my way to the bathroom thinking how fortunate I was to have the whole of the bathroom to myself. Normally there was such a stampede that it was not unusual for as many as six to eight to be looking in one mirror at the same time. One fellow declared that he had lost his razor, but each morning for a whole week had gone down to the bathroom at the height of the rush, pushed his face up from under the crowd as they surged round the small mirror, and had received a perfect shave from either side at the same time.

I had no intention of trying this to see if it worked, but it did cross my mind that we were to be shaved and ducked during the day. Knowing how heavy-handed a sailor is to his own messmates at this kind of treatment, I decided to carry on with my ablutions. Then I found I still had time on my hands so I made for the open waist of the ship to breathe in the fresh air, and then returned to my sleeping messmates to waken them with a freshly made cup of tea.

The young Royal Marine had sounded 'Reveille', which had changed their snores into grunts of woe, as now an arm stretched out, or a foot shot through the strings of the hammock, a muffled oath, a remark 'I wish I had never joined', or 'Why did I leave my dear old home?'. As each one stirred and raised his head to look over the side of his hammock, I offered him his cup of tea.

As this was the day given over to crossing the line we were soon preparing for the great display, with ratings from all parts of the ship giving their help, as did all the officers concerned. The Commander had made his final inspection and the tarpaulins had been rigged into a square shape, filled with sea water. A large platform had been erected with a tip-up chair, and in a very short time all was in readiness as 'Hands to breakfast, Rig of the day bathing suits' came over the loudspeaker. All ship's company not having crossed the line had to muster by the initiation pool from 10.00 hours onwards. Then followed our breakfast of eggs and bacon, the messdeck was cleaned up, bathing trunks taken from the locker and all was ready. We all swarmed over the ship. On the boat deck now appeared all the officials and their attendants, taking up their positions, with King Neptune and his Queen being seated on their two thrones placed high up on the raised platform. All the attendants were dressed in ancient seafaring dress with wigs, pigtails, blackened faces, painted bodies and a press gang of a dozen husky men who chased and caught any man whom they thought needed a ducking. Neptune declared the Court was in session; if a man did not volunteer, he was soon seized and placed in the chair and anointed by a bucket of water, and as he opened his mouth to gasp for air, a barber pushed a paintbrush of soapy lather in it, and shaved him with a wooden sword. Then he was tipped backwards into the bath, where willing hands dipped and ducked and sat on him for about five minutes, before allowing him to scramble out at the other end of the bath. The whole of the ship's company, including the Captain, received this treatment. All were passed through the Royal Court, willing or not it did not matter once the press gang got your trail. This lasted the whole morning, and by that time most of the Court Attendants were tired out and all had been accounted for. King Neptune declared that the ship's company would be treated with the respect due to those who had crossed the Equator, and the ship would be allowed to pass through the Royal Domain. His Royal seal would be stamped on a Certificate and given to each subject. He then declared the Court closed and two large dolphins leaped from the bath, placing themselves in front of the royal throne. The King and Queen mounted their steeds and were conveyed with steady flap of their fins on the wooden deck to the hatch of the cable holder, their route being lined with a guard of honour presenting arms in a royal salute. The last I saw of Neptune was his crown of silver disappearing down that cable holder. I now have a Royal certificate, and this is what it says.

H.M.S. *York*
To whom it may concern

Whereas by our Imperial Condescension – We give this as a Royal Portent under our sign Manual to certify that the undermentioned person has this day visited Our Royal Domain on board His Britannic Majesty's Cruiser YORK and has received the necessary requisite initiation and form necessary to become one of our loyal subjects. Should the undermentioned person fall overboard we recommend all sharks, dolphins, whales etc. to abstain from eating, playing with or otherwise maltreating him. And, We further direct all Sailors, Soldiers, Marines, Globetrotters etc. who have not crossed Our Royal Domain to treat him with that respect due to one who has visited us.

Given at our Court on the
Equator in longitude 30°24'W.
this 22nd day of February 1934
                R.D. SANDERSON A.B.
            Sealed and Signed by Neptune Rex.

This now entitled me to a little more swagger and a more rolling gait in my walk, but as I was still only the lowest rating I could not pass this off, although I had now become an Able Seaman, this being automatic as the course in Barracks was assessed as passing for this rate and only practical sea-time experience was necessary. This being accomplished I was instructed to make a request to see the Captain, through the Commander, through my Divisional Officer, to be rated Able Seaman and to present first to the Leading Hand who would get it signed by the Petty Officer of my Division, returning it to me, and then I had to present it to the Master at Arms' office by 09.00 hours on Monday morning. He in turn read it, then cast his eyes over me and said 'A.B.!', shook his head from side to side, saying, 'I expect you will be rated. They seem to make anyone A.B. these days. Attend Commander's request on Thursday and don't be adrift.'

I had turned up punctually, looking my smartest in my white tropical suit. The Commander had signed the request, and just said 'Captain', without even looking up. I had then doubled away to await another week before attending Captain's request men, which was almost the same procedure with one exception. The Captain looked straight at me and I felt my inside turn over as his piercing eyes seemed to penetrate right through me. I thought of the Master at Arms looking at me and shaking his head, and for a moment doubt gripped me, but my pride came to my rescue. With all the Captain's gold braid and years of command, I could look him straight in the eye, humble as I was. With a flourish of his pen, he looked down and just said 'Granted'. 'On cap, right turn, double march' from the Master at Arms, and I made for my mess. Proud and delighted I straight away rubbed out the notice where my rating appeared as 'Ordinary Seaman, Cook of the Mess' and put in large block letters 'A.B.'. That evening I wrote home to inform my family to address my letters 'Able Seaman' from then on.

All kinds of jobs were taken up by a few of the hard-working boys to earn themselves a few extra shillings, perhaps to send home to their wives, or just to have a beano ashore. There were the photograph firm covering the whole voyage, the barber who would be in attendance during the dog watch, giving us the short back and sides with his patent hair clippers, usually situated amidships where the ship had the least roll, otherwise you got up from his chair rather lop-sided and he told you to come back when the ship was rolling the opposite way so he could straighten his work up. There was the dhobi firm, who would guarantee to have your best collar ironed and starched, all whites washed in 'Persil', hammocks and dirty overalls a speciality – 'you make them dirty, we make them clean' – was their proud advertisement. Their 'Persil whiteness' was mostly ship's hard soap and their speciality was strong soda from the Bosun's store. There was a drying room for this purpose and on certain afternoons lines were allowed to be rigged up on the foc's'le in the hot sun. In the sea breeze, washed clothes dried in a very short time. There was a night for tombola and each week of the football season a sweep. Then, of course, there were one or two who were bookmakers, or the dark horses who always seemed to have lots of money, and you saw them in dark corners with a cluster of men around. Out would come the Crown and Anchor Board, or 'fraze' school in which you placed your money on a certain line of cards, and you won if a clear line turned up before the Joker appeared. This had to be done secretly as gambling was strictly forbidden and if caught would most certainly mean so many days 'chokey' (imprisonment), but it always seemed to me that the more strictly forbidden, the more exciting the game, and I occasionally invested my sixpence or shilling.

One evening, as I sat in the mess darning my socks, one of my shipmates who was from my own district, with a few years of experience as he wore two good conduct badges, hurried into the mess, turned to me and thrust a musical instrument in my arms, so quickly that I pricked my thumb with the darning needle. 'Your name was the first drawn out,' he remarked, 'and you have won the banjo.' What with pricking my thumb, and having a stringed musical instrument thrust upon me, I sat with my mouth open with no sound coming from it, as I think surprise and pain had met halfway down my throat and had got mixed up. He turned round and disappeared quickly. My messmates, only too helpful in a case like this, gave me what was supposed to be a gentle slap on the back for being so lucky, but which was so hearty that I finished up on the deck, with the needle still in my thumb and a couple sitting on top of me. Still being out of breath, all I could do was to withdraw the needle and insert it in a large target where the weight was heaviest. This had the instant effect of releasing the heavy weight.

On rising and looking at my awe-inspiring prize I fingered it gently, and then began to draw on the strings. I had not an atom of musical talent in me at all, so I soon left off and began to wonder how I had won this prize, as I could not remember buying a ticket. So I made enquiries in the mess. I fully expected someone to come and say a mistake had been made and take it from me, but after about half an hour nearly everyone in the mess was telling me how lucky I was and I thought I had better go and find out something. On returning to the mess rather quickly I overheard one of the fellows saying that I had had a ticket, because the owner could remember selling me one. This put me in rather a strange predicament. I had now kept the prize so long. My messmates, through my remarks, had thought I was in some risky business, but now seemed to be satisfied. I thought I had better remain quiet and let things sort themselves out and turn in for the night. Before I fell off to sleep the Leading Hand of the mess came and offered me two pounds if I decided to sell it. So I slept dreaming of the banjo and the pound notes.

In the morning I found out that there was a P.O. with the same name as myself. The man who had sold the ticket had omitted his rating, and the prize had been given to me because I was genuinely thought to be the winner. He said that it would be as well to leave things as they were, and as I had been offered £2 if I sold it, I could keep 30s. for myself, and give him 10s. as they had not made any money out of the draw.

I decided to wait a few days, then sell it to the Leading Seaman, and if all kept silent, there would not be any trouble for anyone. This was duly carried out, I got my £2 for it and gave 10s. to the organiser, I did wonder a little whether he was still on the make, but I never heard any more of this incident.

CHAPTER SIX

A few more days and we would be in Rio. We had left Gibraltar on 21 January and would arrive in Rio on 28 February. I was certainly getting my sea time in, and missing Old England's icy winters. It did not seem to me now that cold weather ever existed. I sometimes thought of home and the country, but not very often, and not for long. Life had changed so rapidly and taken me along with it. 'South America, here we come' was the cry as we lifted our tots of rum and toasted our King. Some added 'God Bless Him' as they smacked their lips and drank the last drop.

I had never expected to see a harbour as beautiful as the one we sailed into one morning. It was as though a lovely picture had come to real life. A picture of mountains rich in deep green foliage, splashed with jagged edges of reddish rock. Dotted about on the coastal slopes were white villas with large verandahs and red-tiled roofs. On one side was a high, slender mountain in the shape of a loaf of bread, which I learned afterwards was called 'Sugar Loaf', and in the other direction was another mountain with a statue of Christ on its very summit. This statue is 90 feet tall, and from its position it overlooks the whole harbour and town for miles around. With his arms outstretched, it has a most commanding appearance. We fired a twenty-one-gun salute on entering the foreign harbour, and it was returned by the shore battery. We now anchored in this wonderful bay, and rigged canvas awnings as the sun was scorching hot. The anchor cable was painted and the ship's side touched up to give a smart appearance, and the whole of the ship's company was reminded that we were all ambassadors of Great Britain, and were expected to conduct ourselves in an orderly manner.

I did not rush ashore, but decided to wait, as in the next day or so a berth was being found for us alongside the sea wall. The enchanting beauty of this harbour held me back, and as soon as the ship had anchored and all were settled down, I climbed up to the bridge and drank in this wonderful view. Then in the midst of one of the most beautiful harbours in the world I let my mind drift back. I left the bridge, quickly returned to my mess, and put pen to paper. I would do my best to convey my thoughts home while I was in this frame of mind.

It was dark when I finished my letter, and as the night was warm a few of us took our hammocks and spread them out on the upper deck, remarking on the glittering lights around us, with the red and green of the ships and boats as they silently moved to and fro about the harbour. Suddenly there was a small gasp of astonishment from one man as he pointed upwards, and behold, high in the sky, as if suspended in space, with the stars twinkling all round, was this great statue of Christ, all aglow in the blackness of the night, with its outstretched arms and face of calm assurance. It seemed to me that it was watching over all who dwelt beneath and guarded our souls from all evil. I felt nearer to heaven than I had done all my life. I said a prayer and fell asleep.

We had now berthed by the sea wall, right in front of a large skyscraper, the first one I had ever seen. I went into this big city, teeming with life. A number of us piled into a taxi which would normally hold five. This time, the driver shoved in more than twice that number. With an effort the door was closed and the windows opened. With a smiling countenance he leapt into the driving seat and with his foot always hard down on the accelerator he was weaving in and out of the traffic, flourishing his arms as he pointed to some building or other, waved to another taxi driver, yelled at someone along the side, sounded his horn every half minute, braked sharply then swirled round a corner. We were saved from bumps and bangs as we were so tightly packed we acted as a cushion for each other, but that did not stop us from being partly suffocated by smoke from his large cigar, every time he turned, still smiling, one hand on the steering wheel, the other removing his cigar, blowing clouds of smoke as he uttered some gleeful remark. We tried to answer back in the same smiling way, but he only seemed to take this as an incitement to go even faster, and we became rather alarmed as a lorry zoomed round the corner and both drivers jammed on their brakes. As we jolted to a stop the door burst open and we tumbled out before the driver could light up again. We paid our fare and were only too glad to carry on walking, to the astonishment of the dumbfounded driver as we left him calling and imploring us to re-enter his taxi.

The South American ladies were beautiful, graceful and elegant, with lovely dresses sweeping down to their ankles as they left their plush-seated cars to mount the steps leading to some grand hotel, and from the large open windows the sound of music drowned the noise of traffic and floated through the night air. We moved further on to get away from the bewitching atmosphere and went deeper into the town where the streets grew darker and closer. A young native stopped and addressed our party, imploring us to follow him. The more we hesitated, the more he insisted with his few words of broken English: 'Come, much dancing, Señorita'. We allowed ourselves to be led to a ramshackle building where the wailing notes of some muffled instrument greeted us as our guide opened the wooden doors, and ushered us in. We looked around the dimly lighted place. An elderly Señora advanced towards us, then smiled and turned towards a table, placing a number of chairs for us to be seated. A waiter quickly produced a bottle of wine and glasses, the three-piece band wailed even louder, and from a curtain of a side partition a young girl of exquisite beauty came gliding on to the ballroom floor. Her eyes flashed under her heavy dark lashes. She had a pleated dress which hung round her ankles, but above her waist was bare, apart from a slender bikini strap. Her hair reached down to her shoulders. She raised her arm and poised herself in a graceful stance, then slowly her arm quavered and the sound of her castanets increased as she began to sway to the music of the band. Her wooden heels beat out rhythmic taps as she began her Spanish dance. She glided round our table exposing her bare leg through her pleated skirt. I turned my head, not knowing what to do, but one of the party with a bit more composure than the rest of us stood up and raised his glass to the fair Señorita.

With a loud shrill prolonged note from the band, and a rapid tap raining to a final stamp, the dance came to an end. 'Well, boys', said our spokesman, 'We must ask the lady over, it's an old Spanish custom.' 'Please don't,' said Shorty, 'it makes me feel embarrassed. Let's go,' and he again rose from his chair. There was no need to ask the dancer, for as Shorty rose from his seat she was right by his elbow, and thinking it was a mark of courtesy on his part, she smiled and drooped her shaded eyelids as she seated herself in her chair, placing her bare leg from her pleated skirt over the other one, right under Shorty's very nose. I think he wished the floor would swallow him, but I lost no time in giving him a nudge, telling him his promptness outdid the lot of us. 'Now follow it up by inviting the Señorita to refresh herself and get her a nice drink.' Much to my surprise he promptly did so.

We all looked at each other, and she at us, wondering where to begin. Her glass was re-filled at our expense and she moved from chair to chair, her feminine charm putting us at our ease as we paid for her ever-emptying glass to be filled. 'Señorita, have you more sisters, friends, girls, more Señoritas?' we asked in our broken English for her to understand, and we pointed to the number of us and counted on our fingers. 'Find us six more pretty Señoritas and a small one for Shorty', and we indicated this by measuring up the poor chap and pressing down with the palm of our hands to his size. She seemed very quick to catch on as she pointed to the clock, and bade us wait by signs as she skipped out of the door.

We all fell to wondering what was going to happen next. Would she really bring back some partners for us?

We were so absorbed in our general chatter that we never noticed the door open, but suddenly we found ourselves besieged by an equal number of the opposite sex, who without any introduction or shyness on their part made straight for us, either seating themselves by our side or standing by our chairs. We eyed them up and down – these ladies descending upon us made us rather backward in coming forward, for it seemed too good to be true. Someone broke the silence by saying 'No wonder Drake always sailed to these parts. Let's have some music.' The band played some dance music and most of us rose to our feet to take the tender little hand of our Señorita. We danced rather tamely at first, but as the music progressed we began to loosen up and become more gay. As the dance finished they held our hands, leading us to our seats, filling our glasses and looking imploringly at us and then at the waiter. There was no mistaking what was expected of us, and a glance was enough for that waiter to be at our service, filling the glass of our partner and bowing to take our money and the tip we offered him. The Señorita squeezed our arm and sidled up more tenderly to our side as each drink was forthcoming. We were getting along quite happily and making ourselves understood more easily than would have been expected, the language barrier seemed to be an advantage instead of a handicap, as with signs and expressive pointing and gesticulating the Señorita gave a little laugh as the meaning of a question was grasped, and then made an answer for us to puzzle out in the same way.

As the evening wore on, our little dance hall began to fill with other young men and our Señoritas were occasionally claimed as partners. We then decided we would move on to see more of Rio night life before finding our way back to the ship. We walked on, wondering which way to go, when up stepped our friend who had conducted us earlier, again ready to press his services upon us. Idly we followed him as he made his way crossing streets, turning corners into what seemed to be a maze of small houses with shaded lights and short verandahs with scantily dressed Señoritas parading their figures. Here we discovered we were in the sordid part of a foreign city. Hastily we looked around, pressing our guide to lead us back to our ship. I looked up into the sky and saw the floodlit figure of Christ with its calm assurance, and arms outstretched as if to take in the whole world that lay beneath him. I said a prayer for him to help put things right, then fell asleep.

I remained on board for the remainder of the stay, except for one afternoon when a party from the ship were invited to go on a tour to the top of the mountain. Miniature trains carried us almost to the top, then we climbed the remainder on foot. From the summit we viewed the whole of the city as gaps in the cloud permitted. For the first time in my life I was standing above the clouds at the foot of the huge statue. Only the very tops of the mountains were visible above this fleecy bed of down, and when a gap appeared the sun sprayed the city with a silver stream of light, which swept over the green forests on the mountainsides to the seashore and out over the sea. I felt elated, and happy to see such a magnificent view, and it brought back childhood memories of the countryside when I had stood on the highest hill back home.

The following day we steamed out of Rio harbour, saying 'Goodbye' to all the wild night life. We were to call at a few places and finish up at Bermuda, our base. Most of these were small coastal towns where we were made most welcome by the few British people who had settled out there. Most of these people seemed to be of Scottish descent. One small town filled a large hall with the choicest of eats and barrels of beer from which to help ourselves. At another coastal bay where forest swept down to the seashore we were allowed ashore for recreation during the afternoon, to enjoy the company of natives who lived in small reed huts and went fishing in boats hollowed out of large tree trunks which they rolled on logs of wood to enter the sea and bring them inshore again. They allowed us to launch them on their rollers into the surf which broke over this part of the coast. At first it always capsized and threw us in the sea, but after several tries and with their help, we were successful in keeping them upright. Their natural, happy laughter made a very enjoyable afternoon.

We steamed up the mighty Amazon, and saw large reptiles lying on its surface and circling our ship. We also visited those far-flung posts of the British Empire, the islands of Trinidad and Barbados. Here the natives would paddle out in their small boats. Some would be allowed to come inboard to sell us coconuts and fruits of all kinds. Others would stay by the ship and dive for the coppers we threw down to them, and for sixpence would dive and pass right under the ship to come up the other side. At Barbados I celebrated my 21st birthday. On dropping anchor I was ready waiting for the first liberty boat ashore. I celebrated by having a nice chicken dinner, then made friends with some Canadians who were from a sailing schooner in the bay. We toasted our countries with Barbados rum into the early hours of the morning, then fell asleep on the chairs we coupled together. A black girl woke me up in the morning by splashing my face with ice-cold water, telling me to go back to the ship, as it was 7 o'clock. From the sound of the snores of my companions they seemed to be filling their sails with wind and sailing on a sea of rum judging by the smell. I left them flat out and slipped away as quickly as my dazed head and heavy legs would allow me.

Then on to Bermuda, with its naval dockyard, its dry dock for all repairs to be carried out. This was to be our base to return to after each voyage until our commission had been completed. Some evenings a group of us would gather in its large canteen and fill a table with many bottles of good strong English beer, and give a rousing cheer as some gifted sailor would mount the stage to give a turn and sing a song. We would all join in the chorus together and then roll back to the ship full of the warm effects of the life-giving ale.

Our stay here was to be from six to eight weeks. Games of all kinds were arranged for us to take part in. This part of Bermuda was made up of small hills all linked together, dotted over with palm trees and shrubbery, with a road running up and down the middle. No cars were, I understood, allowed on these coral islands. The climate was perfect and the sea around was filled with shoals of fish. Swimming before breakfast each morning from the ship's side was most refreshing. I played cricket and football and joined in long-distance running and there were always boat pulling, boat sailing and plenty of water polo to keep one fit and well. I did well at cricket and reached the last forty in the four-mile road race, before sailing on our next voyage which was to take us through the Panama Canal into the Pacific Ocean.

Our first stop was at La Guarda, a port in Venezuela, and after a perilous train journey I visited the capital, Caracas, situated high up in the mountains. The railway track must have been cut out from the mountainside, and as I looked from the window I saw on one side a sheer drop of hundreds of feet, and on the other, perpendicular rock, so I kept my gaze within the carriage and saw a label 'Made in Birmingham', and I felt more secure to see this familiar name in such a remote part of the world.

Our next port was Cartagena in Colombia, where the British ships in the old days had many a brush with the Spaniards, and Drake had sailed in and sacked the town. Then we sailed from the Atlantic Ocean into the Pacific Ocean through the Panama Canal. Passing through the locks, it seemed to me as if the ship was lifted over the top of a hill and set down on the other side to proceed on its way. We sailed down the coasts of Ecuador and Peru, paying a visit to some oil wells which were managed by the British. There we were made most welcome, the ladies laying on a nice tea and attending to us like Mother at home, making small talk, asking if we had enough to eat and pressing us to have a little more – it made quite a change from the mess table. Thousands of miles from home we found a real bit of Old England in a few grand ladies with their genuine concern for our welfare, the sincerity of their simple questions and advice given us to take care – given as if we were their own sons. They even asked us if we had any old socks with holes in that needed darning, but we threatened anyone who produced any that he would be thrown overboard. However, we did allow Shorty to have the trousers stitched which he had torn climbing ashore, and it pleased the ladies so much to do something for us that next day we brought a collar to be sewn up or a button to be sewn on. They would ask us about our families, how long we had been away, what part of England we lived in, etc. We talked and put on our best manners and sat with them drinking cups of tea, although they had plenty of beer for us to drink.

Our ship's football team played their native boys and our countrymen told us to be sure and beat them. As they came out on to the field, led by their captain, in single file they ran round it, then up to the centre spot and presented our captain with a large bouquet of flowers. To receive a large bouquet of flowers would make any sailor feel ill at ease, especially standing in the middle of a football pitch with all the ship's company cheering him. To the relief of the embarrassed Captain someone from the touchline ran across to take them from him, and let the game proceed, but alas not in our favour. The natives in their bare feet were much too fast for our team, and those few English people had to put up with the boastful chuckles of these black boys for a long time after we left.

The land of the Incas – Peru – with its mountains full of gold where the Spaniards loaded their ships to be carried home to their Motherland was our next stop. The Incas had built up a civilisation far in advance of any other Indian tribe, but found themselves no match for the firebrands and cruelty of the Spaniards. These Indians, in pathetic wretchedness, now haunted the streets of Lima, the capital city. Hopelessness was stamped on the faces of these once proud people. In the hustle of the busy city they wandered around. I often wondered if they would ever regain some of their past glory, or whether they would die out altogether.

One bright morning a party from the ship caught a train from the port of Calloa for a journey into the Andes. Hour after hour the train puffed its way round and round the mountainside, gradually climbing. Mountain streams and roads wound parallel with each other through the valleys at the foot of these high mountains. As we climbed ever upwards, the air was getting much thinner, making it rather uncomfortable. The journey took from six in the morning till two in the afternoon, when we alighted to wait by the track for a train back. We were twelve thousand feet high, and all around were mountains and more mountains. The train, we were told, carried on for another two days. Beside the track where we alighted lay a large number of sacks. Wondering what was inside and with our sailor's knives ever ready to be used, we wasted no time in cutting a small opening to satisfy our curiosity. We found that these sacks contained silver and gold ore. We could see the silver and gold glittering in veins in small lumps of this ore. I took a small piece to carry home and thought to myself, I will go into my local pub on my return, take this from my pocket to show to the village men and boys – gold and silver ore I had picked up in the heart of the mighty Andes.

As we were inspecting our find, some native women (who it seemed lived in some small houses close by), approached us and by their talk and gesticulations seemed to have something particular to show us. On offering them a small amount of money, they quickly returned to their homes to bring with them an old, wrinkled lady carrying a small cage which she held out for us to inspect. Inside this glass cage was the mummified figure of a man, shrunk to the size of a small child. This I understood was a skilled art, known only to these Indians, the secret being handed down in the family.

Down the track came our train to take us back. We waved farewell to the few natives and wondered what comfort they got out of life. The barefaced rock slipped by, the valleys came rushing to meet us. I felt tired and dozed off and awoke to find the train had left the mountain descent and was speeding through the few miles of cultivated land that lay between the mountains and the sea, and soon I was back on board with another memory stored away for the rest of my life.

'Coming ashore? It's the last night. We leave tomorrow.' 'I'll think it over,' I said. Then I turned quickly and ran down the steps on to the mess deck. 'Wait for me. I'm coming with you.' Quickly I changed into my going-ashore suit for our last evening in Peru. There was the Leading Hand who was as strong as an ox, and who always seemed amused when he asked me about my country life; he had twinkling blue eyes deep set in his rugged face, and he became a friend of mine from the commencement of the voyage. I could never understand whether he was laughing at me or not though.

On reaching the landing stage we jumped out, and told the coxswain of the boat not to be too early in the morning. We pushed our hats back and stepped off into the nearest bar, and drank round after round of drinks, until I began to wonder how long I could keep it up. I struggled to keep up the pace, but my cheeks were glowing. I remembered to keep a grip on myself and play wisely. 'I must have something to eat, and I'm going to get it.'

With that remark I walked out into the street with a steady tread, and kept walking for about an hour, then returned to the same bar, to find them all still steadily drinking.

'Who's coming with me? I know a bachelor club to go to,' said our Leading Hand. 'I'll come,' I said. No one else wished to depart from their drinking den, so we left the party to search the streets for a taxi to take us to this club. Finding one at this late hour took us some time, but the biggest problem was trying to make the driver understand that two sailors wished to be taken to a bachelors' club. In vain we kept repeating 'No Señoritas, only Señors', but he looked doubtfully at us, shaking his head as if trying to tell us that no club like that existed in this part of the world. The only way to find it, we thought, was to go in turn to each of the entertainment clubs. The poor driver raced from place to place, we would look at the name and shake our heads, saying 'Not the right name' and wave him to carry on. 'We will try one more, and then give it up,' we said as the taxi swung into an avenue of trees. There it was, half hidden by the trees, the name 'Bachelor' in lighted golden letters in a half-circle over the entrance, giving it the appearance of being a rather more expensive club than we could afford, but we thanked the driver anyway, giving him an extra tip for the trouble we had given him in finding the place, turned and walked up the broad pathway and gave a sharp tap on the door.

It was opened rather cautiously by a steward who ran his eye speculatively over us. We announced ourselves as 'Two bachelors from His Majesty's Ship *York*'. We were ushered into a large room with soft cushions decorating the large plush divan and chairs, with shaded lights, and trees planted in large tubs, with lighted birds in the branches. Looking up to the high roof of the room we saw a shining moon appearing from behind a large cloud with twinkling stars dotted between white fluffy clouds. Across the room, seated at a piano, was a young Señorita, with a small group of others close by. As we entered and the steward carried off our hats, one of them detached herself from the group and slipped across the room to us. Her silken dress rustled as it swept over the floor. She wore flashing jewelled earrings, and sparkling stones round her neck and on her arms. She spoke softly, her voice sounding so musical to our ears, and the perfume that enveloped the atmosphere made my knees feel as if they wanted to buckle under me. The only word I could understand was 'Señor' as she patted the divan and ruffled up the cushions, and then stepped back. I seated myself, and wondered where I was sinking to in the soft cushions – a change from the hard mess stools. My companion took the situation more calmly than I did. He was much more composed, and taking out his silver cigarette case, as if to show the lady that she did not have the monopoly of all that glittered, he gravely offered her an English cigarette, ordered a bottle of wine, and then seated himself very carefully while I was still fighting my way to the surface from the depth of the cushions. I eyed the moon and stars above me, and felt as though I was in a dream. 'How much money have you got, my friend?' he asked, as I tried to shake off the effect this atmosphere was having upon me. 'About £2 10s. in English money,' I whispered. 'Good. I have about £4. This looks rather an expensive place, and we must live up to it.' 'Yes,' I said as my eyes began to open a little wider, as the Señorita was not only bringing the drinks back, but another dark-haired companion as well. I must do something, I thought, and not leave everything to my companion. Gallantly I struggled to my feet. 'Allow me,' said I, taking the silver tray from her and noticing her shapely hand with painted fingernails, 'Please be seated, and would the Señoritas share our table with us?' Without waiting for their consent, I summoned the steward for further glasses with wine for our lady friends. I then turned to look at our new acquaintances, who had now seated themselves on the long divan, the tallest next to the Leading Hand, leaving for me the one whose shoulder straps seemed to be slipping. Her dress crinkled as she moved, and her beads made a faint beating noise as she also sank into those soft cushions. I felt awkward, and almost spilled the wine, but saved the situation by placing the glasses on the table, not trusting my trembling hands. I think my dark-haired Señorita must have sized me up rather quickly, as she reached over to give me my drink, and smiled at me over her glass as she daintily sipped hers. I almost choked as I swallowed hastily, and this made my eyes water. Out came her little handkerchief. Reaching over me, her dainty little hands and small handkerchief were going over my face, so tenderly that I choked some more. She finally wiped my face, put my collar and silk straight, and from her powder compact was dabbing my face. 'English Sailor no powder,' I said, in a half-hearted pretence at pushing her arms away.

We were getting along quite well, in no time at all, and then the piano started playing. Springing up she pulled me to my feet, not knowing whether I could dance or not. In such a lovely setting and with a girl who blended so beautifully with it leading me on, we danced to the music in and out amongst the shaded trees. The piano stopped and we strolled back to the divan. I had even forgotten about my companion until he tapped me on the shoulder to remind me that he was still there, and it was my turn to summon the waiter to replenish the silver tray. 'I don't want to get drunk with wine,' I said, 'this Señorita is much more intoxicating, and she sparkles more than the wine.'

Just then there was a loud clatter at the door and through it appeared two young men pushing the steward on one side, throwing their arms around and shouting excitedly as they entered the room. Until then we had been the only two bachelors in the room, and these intruders were, to me, most unwelcome.

We sat on the divan, my resentment growing, and my Leading Hand's frown darkening, with a steely glint gathering in that seadog expression as our table was toppled and kicked by these two Peruvians in their attempt to dance and whistle in a possessive way with these Señoritas, who in their half-nervous way had to endure their ordeal, and were frightened to protest too openly. Now our wine went flying, and my companion could take no more. In a moment he was on his feet and with three rapid strides, huge hands outstretched (it was a surprise to see such alertness in one so heavy), he had the Peruvian by his neck and pants and piloted him up the middle of the room to the door. With a mighty heave he was bundled through the door and I heard a swish and thump on the gravel, telling me the job had been completed.

While this was going on I too had leapt to my feet, keeping an eye on the other fellow who now, as he saw his fellow partner being so powerfully bundled out of the door, took up a threatening attitude, but kept his distance, much to my satisfaction. I can usually stand up to anyone who threatens at a distance, but the next move almost made me step back as from his pocket he produced a revolver, waving it wildly in the air, which silenced the girls and made me eye him rather more carefully. I did not retreat, not because I was brave, but I think the unexpected, coupled with surprise, held me rooted to the spot and made me unable to move either backwards or forwards. This, I think, made the other chap think he had some strange fearless opponent to deal with (as not even a firearm waved at him made the slightest diference), and he replaced the revolver in his pocket. I automatically released the air in my lungs, as I am sure I had stopped breathing for a full minute, and I began to turn round as still threatening he circled round me. The minute of silence was now over and there were loud screams and panic amongst the girls as they gathered at the farthest end of the room.

Fortunately for everyone the Leading Hand returned, not knowing how we had been getting on, but seeing the situation he immediately went into action and his display of strength as he picked up his second opponent and hoisted him above his head, held him aloft for the Señoritas to see, silenced their screams and called forth instead cries of admiration as he carried him aloft out into the night and pitched him on the gravel path to roll over, just as if he were pitching a barrel on an everyday job. Two bachelor sailors were now the admiration of six or seven young Peruvian Señoritas, who flocked around us, holding glasses of wine for us to drink as another embraced us and planted sweet kisses of hero worship on us. I didn't think I had done very much towards it myself, but who is there amongst us that does not like to be admired? It made my head swim with delight as I realised they looked at me almost as proudly as at the Leading Hand, and I stifled any doubts that entered my head by telling myself that I didn't run away when he waved the revolver.

With the girls all over us and the Steward giving us a free bottle of wine I was enjoying the greatest possible pleasure of my life, but time was passing and as I managed to say a word or two to my companion from the embracing females and soft cushions, he reminded me we ought to be leaving, not only because time was passing, but because the two he had thrown out might be returning with reinforcements, or they might send the police round on some pretext. Seeing the sense of this and wishing to leave as heroes, we shook off our clinging flowers of Peru and they walked with us to the street, and all stood at the gateway waving farewell and blowing kisses as we departed. We turned and waved back until they were out of sight.

It only now remained for us to pick up a taxi and get back to the jetty to await the first boat to take us back. Perhaps we would snatch an hour or so's sleep by the landing stage as the nights were warm, and with a hat for a pillow and no fear of missing the boat, an hour's sleep was the most natural thing for a sailor. About two hours later in the morning, I stood on the deck. We were once again on our way, and the Andes were fading from sight. I was telling my shipmates of the exciting adventure of the night's shore leave. Only a white speck here and there could now be picked out from the coastline. I had some feeling of pity for the Incas. The mountains were to be respected, and the whole country held a deep mysterious aloofness which I still feel to this day.

CHAPTER SEVEN

The effects of a couple of days shore leave had been shaken off, a look at the chart to see our position and our next run ashore which was Iquique in Chile, then to Antofagasta further along, and finally Valparaiso, the biggest seaport on the long Chilean coast, the main port to the capital city Santiago. Here we were going to get 48 hours' leave every watch, so the notice informed us. Thank goodness we get another pay day before the next port of call. Nearly everyone seemed to be broke, for during the dog watches pens would be busy writing home, others caught up on their mending, and the general conversation began to turn towards that 48 hours' leave as we settled down to our sea routine.

About every three months a complete change around of jobs would be given to all the ship's company. Not only as training, but to keep everyone interested and fresh in the running and working of each department. As our attention was called to the noticeboard, I found myself having the job of messman to the Chief Petty Officers' Mess. This excused me from a large number of other duties, being almost a full-time job. I had to join their mess, draw their meals and do all the general cleaning up. The mess held about forty chiefs; most of the real chiefs had joined as boys and risen to the senior rating on the lower deck and were due to retire on pension after twenty-two years' service. Others, such as shipwrights and electrical artificers, had joined the navy as special tradesmen and had been thoroughly trained as specialists in their respective trades. There were three messmen to forty chiefs, a three-badge A.B. in charge who was responsible to the president of the mess for the general cleanliness and orderly conduct of one other A.B. who had one good conduct badge, and myself, as yet I had no badges, just a bare arm and was regarded as the 'spud' boy. While the three-badge A.B. each morning just before eleven put on his clean flannel and white blancoed hat, with, I thought, a certain amount of swagger, to draw the chief's 'neater', I, in my dirty overalls, was just finishing scrubbing the messdeck on my hands and knees with the sweat running down my face.

The other one-badge A.B. informed me he was married. 'Oh dear,' I said, 'to be married to someone thousands of miles away with all these lovely Señoritas at each port. What do you do when you go ashore?' 'I don't very often go. I save my money up, what little I draw, and send it home to the wife. I hope to pass for Leading Hand and then go on for a higher gunnery rating. Each night I write a number of pages home to the wife.' 'You must be a Saint,' said I. 'You're still very young,' he said, 'I used to be much worse and when I drew my money, off I would go boozing every night. It's all finished with. I've got the best girl in the world for a wife.'

I felt rather drawn to his sincerity and faithfulness, and the trust he placed in his wife. 'The next place we call at, you'll have to have a run ashore for some recreation. You want some kind of change, being cooped up in this pantry. Already I'm getting prickly heat all over my back. What can I do for it?' 'Nothing much, use some talcum powder and stop scratching it.'

I became rather attached to this young A.B., and many a friendly talk we had together while peeling the spuds or washing up.

'I shall be going ashore this afternoon or as soon as we have anchored and leave starts,' said the senior messman, 'What again?' said I 'What about us?' 'Look son,' he said, 'when you get some seatime in you can have some leave.' Later on, after we had anchored and the senior hand had gone ashore, I persuaded the other young A.B. to do likewise as I could manage, and leave was only up to 24.00 hours. He prepared to catch the liberty boat, leaving me to do all the donkeywork. Being young and energetic I set about my task with a certain amount of zeal. So intent was I on my work that I was completely unaware of what was going on around me. It was going off well and I was feeling satisfied with my efforts, when to my surprise the curtains dividing the entrance to the mess were pushed aside and my name was called out in rather a loud and hurried tone. 'Yes, that's me. What do you want?' I said, as a boy messenger stood before me. 'Are you deaf or plain dumb? Haven't you heard your name being called over the loudspeaker?' 'No, who wants me?' 'The Petty Officer of the day. Get in the rig of the day and fall in on the quarterdeck, and don't forget your hat and be quick about it.' I changed into my whites, put on my hat and ran on to the quarterdeck where another half-dozen hands were lined up facing the Petty Officer who looked anything but pleasant. and bellowed forth: 'Fall in here. Holding everyone up. Where were you when both watches of the hands were required to hoist up boats?' 'Never heard it,' I answered. 'No excuse,' he replied, 'Stand to attention.' He then walked over to the officer of the day, who was pacing up and down with his telescope tucked under his arm as if he was Nelson's lieutenant. The P.O. saluted, saying something to him and the officer looked at his watch and then across at us, then at the gangway as if he would attend to us in due course. Someone was coming on board. Meanwhile we could remain standing at attention while he bided his time. There we stood like six dummies, not daring to move as the P.O. walked leisurely around, no doubt pleased with our discomfort. Out of the corner of our mouths we hurled some terrible insults when his back was turned, including what we hoped would happen to him in his future career. At the slightest move on our part he would shout 'Stand still'. As our aching legs began to quiver he relented and stood us at ease. In due course the Lieutenant was ready to deal with us, and we were called to attention, off caps, and charged with being absent from place of duty and all asked what excuse we had.

We all seemed to give the same answer, 'Never heard it.' From the P.O. 'On caps, Commander report, Double March, Break Off.' Now furious and bemoaning my fate, I went back to my duties and silently turned in my hammock with 'Commander's report' hanging over my sleeping head. Until now I had kept my conduct sheet clean and I had never been in any trouble as I could never see any sense in losing leave and having to double round with a rifle held above the head, or to do the dirty work which had been left for the men under punishment to finish off.

In the morning I informed my two workmates of what had taken place. They listened attentively, the senior one suggesting that I would incur a heavy punishment. For the next day or two I was very quiet and silently went about my work. I was sent for by the Master at Arms and told to attend defaulters at 10.00 hours Thursday morning. As the day approached, the senior hand detailed me to have myself all ready for a mess change as I would have to change my job to do my punishment. 'How do you know what I shall get?' 'There is nothing so sure. You haven't got a leg to stand on.' I felt I should have to make the best of a bad job, but I was never one to be defeated.

At Commander's defaulters, my own Divisional Officer looked down the ranks at us insignificant ones as the superior beings in their gold braid paced up and down, carrying on their conversation as we stood in silence. Having recognised me, he came over and enquired why I was here, and which officer had put me in Commander's report. On being told it was the lieutenant of the quarterdeck who was Duty Officer that day, he told me to think up some explanation. 'Say something, say anything, speak up,' was his hurried final word. This small word of encouragement in my distress made me always respect him afterwards. I heard later that he had always been at loggerheads with this rival officer.

With great pomp and ceremony, decked in splendour in his spotless whites, gleaming epaulettes and golden laurel cap, the Commander mounted the rostrum, with his Lieutenants, Sub-Lieutenants and Midshipmen all round like foxhounds following their master, the Master at Arms assisted by his Regulating Petty Officer barked out 'Commander's defaulters' (with an extra emphasis on 'Defaulters' as though we were the whole cause of all the trouble in the Navy); 'All present.'

First offender was the young Marine bugler who had sounded the wrong call at the wrong time, he had forgotten it was Sunday, depriving the ship's company of an extra half hour's lay-in. Seven days number eleven. The duty cook was charged with insufficient attention given to cooking the ship's company's food in allowing number ten mess potatoes to boil into a pulp. Three days. Two A.B.s who had overstayed their leave by two hours made the excuse that they had not altered their wrist watches and were still going by Greenwich Mean Time. This made the Commander blink his eyes and ask those around him whether Greenwich Mean Time would be in front or behind the present time. As none of them had the correct knowledge 'Stand over to consult the Navigating Officer.'

My turn was creeping nearer, and my brain was working overtime to think of something to say. I hadn't years of experience behind me and this was the first time I had been a defaulter; I knew that on my report on leaving the barracks was written 'Gives no trouble'. Now my lips were parched and my throat dry, I heard my name and I tried to spring to attention, 'Double March.' The order sounded like a big gong as I doubled up before the Commander and saluted. 'Off cap,' came the next order, then followed my name and number. 'Absent from place of duty, namely both Watches of the hands at 06.00 on such and such a date.' I looked up at the Commander whilst this was being read and realised what an advantage he had over me in his desk and stand, which were two feet above me. His aides who were clustered all around were staring straight at me, while the jaunty large figure with an enormous voice dared me to deny it or say a word in defence. I felt a poor forlorn figure, and as if a whale was about to swallow me. Now the Commander was opening his mouth and looking down at me as if he could see right through me, 'What have you got to say?' Suddenly something in me rebelled and stung me into some kind of retaliation. Possibly I resented being looked down on. 'Well, Sir,' I drew a deep breath to get my wind to carry on with my defence. 'I was alone in that mess to look after forty Chiefs. The senior hand had gone ashore and the other was ready to go which left just me, my hands were full with trays of food and the tea was getting cold. I had no time to obey the call for I was left to do the work that two others always did . . .' I could not go on any further as my deep breath had run out and I had no more air. I looked the Commander straight in the eye with a truer look than his own, and I saw his mouth still open as if he was taking a deep breath as well, and he was quite taken aback, I think, at my short sharp reply. 'Case dismissed. See this rating is not left again on his own. Only one leave at a time.' 'Very good, Sir,' replied the Master at Arms and I thought the jaunty voice was a little less jaunty as he called out 'Case dismissed. On cap. Right turn. Double march.' I nearly tumbled over. I felt quite bewildered, my brain was saying 'Dismissed, dismissed.' I suddenly realised this meant I had got off. I started back for the mess, my feelings changing and a smile showing on my face that a short while ago had looked so tense with worry. Both my working chums were waiting for me, and I tried to keep calm and hold down my excitement. 'How many days?' I left them to do the questioning. 'What days? Case dismissed,' was my smiling reply, 'and only one ashore at a time. It is a wonder you weren't run in for leaving me to do all the work.'

As the opportunity presented itself, full facilities were provided to keep one fit and well in sports. There were teams for water polo, basketball, boxing, football and cricket. and we were often called upon to take on the best that the small ports, as well as the larger ones, could put up against our men. With much work, skill and enthusiasm a Tattoo was given, at each port of call down the Chilean coast. The money taken was to be given to charity. At the final display at Valparaiso our boys as a team had danced the hornpipe, our physical training instructor with a number of muscular-bodied young men had put on a gymnastic display, while the torpedo branch with their electrical skill had given in miniature form a night encounter between warships with big guns firing, torpedoes exploding with flashes and big bangs to make it as realistic as possible. All those with any talent to display gave their help to make it a great success.

Before one of these shows which was to take place in the evening, a free invitation to attend a race meeting was given to all the ship's company, and this appealed to me. I caught an early Liberty boat.

At the racecourse, I was taken to a cheap entrance. As I looked around I could see the place was all wired in. There were no seats at all, and very little of the race was to be seen. As I suppose I was the only British sailor in that particular entrance I had a jabbering crowd of racing folks all pressing to give me the winner of the next race. They all seemed happy and smiling in their excitement, declaring I ought to back this one or that as they pointed it out to me on their race card. I thought I might as well join in with them and smiled back. Anyway, it pleased me to be the centre of attraction, even if it was only in a third-class ring of a South American meeting. So I bought myself a beer, tilted my cap back, and smiled back at my fans. I scanned their race card, shaking my head at their suggestion, and picked one out for myself. 'No, no!' they seemed to warn me, but waving their protests away, I proceeded to the Totalisator to place my bet. I could not pronounce the name, so the number was the only thing we had in common. Then we all went to the wire which separated us from the course to await the start. I couldn't see much, but the roar from the crowd told me they were off, and I stood there calmly, watching the crowd encouraging their respective horses with shouts and cries. As the excitement grew louder I managed to catch a glimpse, and saw my number coming along. With a tremendous burst of speed it sailed into the lead, and I felt like throwing my hat into the air, but checked myself to show my

British coolness. A number of these young fellows were turning to me and leading me to the winning box to collect my money. I didn't really know if the horse had won, but they were all pushing me to the pay-out box, and after waiting for about five minutes the shutter went up, and I gave in my card and was paid a handful of paper notes. For the next race the same procedure took place again, with my followers pointing out which I should put my money on. Now laughing more confidently and brushing their choice to one side, I picked my own. Again it came up, and the smiling crowd pushed me along to get my winnings.

Even the man at the pay-out box was all smiles as he watched me slipping the notes in the purse attached to my belt. Altogether I had five bets, and four of them won, so I spent a most enjoyable afternoon, with the crowd enjoying my luck with me. At the finish they were giving me their cards to mark. I had won 200 pesos. I caught a bus back to town to get rid of my racing fans, as I was now beginning to find them rather embarrassing. I then had something to eat and a drink, and went back on board with a good sum by me to help my expenses on my leave at Valparaiso, to which the whole ship's company were now looking forward.

The rattle of the cable passing through the cable holder was the sign that the ship was anchoring at Valparaiso. One or two men dashed away to have a look at their surroundings, while others prepared the breakfast and passed their witty remarks on to each other as to how they would spend their leave. In no time at all the mess was cleared. We assembled on the quarterdeck as if we knew what was coming. The Commander said a few words to the sea of upturned, expectant faces. A berth was being given us by the jetty, and we would be weighing anchor and proceeding alongside by noon. There would be shore leave up to midnight. Then on the following day, starboard watch would proceed on 48 hours' leave at noon, followed by the port watch.

Soon, the anchor was weighed, and with all due pomp and glory in our best white suits, we cleared lower deck to line up in ranks with the tallest for'ard and the shortest aft, each in their respective parts of the ship, with the marine guard and band on the quarter deck. At last *York* was safely secured to the dockside. We had all done our job well and the Commander's beaming smile was enough to convince us that he was satisfied with his crew, and he could report to the captain that his ship's company was in good heart, and was just raring to take the town by storm.

The crackle of the speaker was heard, and even before any audible sound came over the broadcasting system, a stream of libertymen were filing out of the messdeck on to the open deck, piling down the gangway to muster on the jetty in their smart attire, with that extra eager expression which coloured their whole bearing as they stood to attention listening to the last warning of the Master at Arms of the final time of expiration of their leave. With the last 'Right turn, quick march, carry on libertymen', a surging mass of sailors were at last making for the open streets, all bent on whatever came their way.

CHAPTER EIGHT

With the excited hum from everyone talking at once, I made my way with the rest to enter the streets of Valparaiso. First of all some of us decided we would change our money. We boldly stepped in and smacked down a pound note, to be given twelve notes back to the value of ten pesos each. These twelve notes, I was to discover, went as far as twelve pounds would have done in England. Down the street we walked, eyeing the shop windows and watching the busy life of the town go by, much the same as any other town at home apart from the language and the continuous honk of the motor horn. Once ashore, it is not long before Jack begins to feel hungry. I suppose it is not altogether due to the ever-growing necessity to eat, but a lot to do with it is the yearning to see that tablecloth and cups of steaming tea, and to be waited upon like Mother does at home. It was not long, therefore, before some of us were stepping through the glass doors of a first-class hotel, with the confident air of security which those twelve ten-peso notes gave us to start off with. Our hats were taken care of, chairs were held for us to sit on, while another waiter entered, uncorked a bottle of red wine and filled our glasses.

Now the waiter returned with folded napkins and covered dishes. Our first course. He attentively arranged everything in order, and we tried to impress him with an air of respectability and behave as if we were used to being waited on. We were rather taken aback though when the dish covers were whipped away and the dishes were replaced by more and more food, and we eventually began to wonder when it was all going to end. Altogether we had a nine-course dinner. It came to a finish with black coffee and two bottles of wine, that washed the contents of the plates down. Some very satisfied sailors were now settled back, smoking a large cigar as they viewed their life through clouds of cigar smoke, and thought at that moment that this life was a most delightful and pleasant one. With a show of importance we beckoned the waiter. 'We would like our bill.' 'The hour of reckoning', as my friend called it. The waiter bowed very politely as he went out of the door, to return in a few minutes with a silver tray upon which was the bill. We examined it in a 'couldn't care less' attitude and wondered what all those figures meant. We covered it with one of the ten-peso notes, hoping for the best but fully expecting to be asked for a few more. Our friend the waiter murmured something and vanished again, to return with the tray containing some notes and a number of coins, which made us all look at each other with surprised wonder that our ten-peso note had paid for all our grand first-class dinner and left a heap of change as well.

Delighted, now realising we were so wealthy, we tipped the waiter with two notes from the change and rose to our feet. He in turn seemed very pleased with our small tip and he stepped silently and sedately to a shelf in the room, picked up a brush and gave us and our hats a brush down.

Now that we had eaten, and it was still quite early in the afternoon, and it was much too hot to do a lot of walking, a visit to one of the large cinemas was as good a way of spending the afternoon as any, and 'a way to pass the time until the night clubs open' was my friend's remark. It was an American film with English soundtrack and down either side was a version written in two or three languages. On finding ourselves once again on the street in the cool evening air, the rush and noise of the surging life of its people had greatly diminished, and only a trickle of the crowd was left on the pavements. The clattering buses and cars with their endless honking and hooting had nearly all left the streets, and only at long intervals did one hear the rattle of a bus or a car swish by, and it made a nice change for us to walk down the long streets and look at the lighted windows.

We drifted idly along the street, without a care, and very soon there were seven or eight in number who had passed the afternoon away in similar manner to ourselves, and we now collected together and looked round for something to attract us like moths attracted to a flame. A flashing neon-lighted entrance in a dimly lit street is at once an invitation or a challenge for a sailor not to pass by. Boldly we accepted this challenge and entered through glass doors, then further along across a compartment into a passage where a larger entrance hall with a double door was opened for us to enter. We found ourselves in a spacious hall. The centre was brilliantly lit, with a square, partitioned dance floor, complete with band. All around the sides were shaded alcoves with tables and chairs, partly occupied with men in evening dress and ladies in evening gowns. Soft carpeted floors drowned the sound of our footsteps as we were shown to our tables. It was very quiet with murmured conversation being carried on by the guests already seated and dining. We nodded our approval as white-coated attendants seemed to be asking us something. Soon our table was filled with food and drink, and we ate, drank and talked and took a look at our surroundings. We pondered amongst ourselves whether these gentlemen in evening dress were businessmen taking out their wives or girlfriends or secretaries. We also noticed quite a number of young ladies unchaperoned, beautifully dressed, reclining on cushioned seats all round the hall itself. No doubt waiting for their male escorts we thought. It was quite early in the evening.

As we watched, the tables were only now beginning to be fully occupied, and the bandsmen had just settled in their seats and were putting their instruments into position and making sure that all was ready for their evening playing as they blew a note or strummed in final preparation. The quiet hum of conversation had risen slightly as the number of members increased, and more of the reclining ladies seemed to be present round the hall. Quite suddenly the lights changed from a white brilliance to a shaded blue, and softly the music from the band broke over the hum of voices. A few couples left their seats, the men leading their partners on to the floor to glide to the soft beat of the music.

We saw with astonishment that two of our shipmates were dancing, holding their partners with all the self-assurance of perfect ballroom dancers. As the dance ended we loudly joined in the applause in honour of our shipmates, who now on their way back were making their way over to our tables. 'What about an introduction?' most of us shouted in chorus. 'Introduction!' said Lofty, 'Are you blighters just dumb? All those girls sitting round the outskirts are your hostesses for the evening, whichever one is your choice. Go and ask her, or get one of the attendants to fetch her over for you. Buy her drinks and you have a partner for the whole evening.'

We listened no more, as the whole bunch of us raised our heads and eyed the young ladies, their figures, faces, dresses and size. All came under our searching gaze, 'I feel like a Sultan picking his harem', was one remark, whilst the thinnest one amongst us preferred the plumpest girl, while the stout one said: 'If you have that plump one, I'm going to have that thin one, she needs a good meal by her looks', and off he went to claim his hostess for the evening, to return immediately hand in hand. We placed a seat for her to join us. Now the ice was broken others soon turned up with their choice for the evening. One or two hung back for a while, whether from embarrassment or whether they were waiting for some divine beauty to appear, or perhaps they were over-cautious and did not want to rush head first into the waiting arms of the reclining beauties of Valparaiso, but soon each one of us had a hostess. No interpreter is needed to transmit or translate that touch or look that speaks so clearly to you both as the evening passes. The wine flows freely, and through blurred and hazy vision your Señorita grows ever prettier. By now you have forgotten completely about your chums and only have eyes for her, and you find yourself no longer at the table but sitting quite serenely together in the alcove, and as the lights dim she lays her head on your shoulder. You feel her soft hair touching your cheek and drawing her closer you slightly bend your head to kiss those lovely South American lips that you feel so sure cannot resist you, but oh so cleverly, up would come that little fan to check an amorous sailor. As you manoeuvred for another advance on would come the lights, and out would go that little hand holding her glass and turning her head to look up into your eyes, speaking some words that you didn't understand but knew just what they meant. The waiter came up and you fumbled in your pocket for some money to pay, and watched each time as he would then turn to her and pass some small chip or something to keep count.

Swiftly the night passed on and into the early hours of the morning, the place was beginning to empty. 'I must take my Señorita home,' I said, 'It's an old British custom. Me take you home to sleep', and I cupped my hands under my chin and closed my eyes. She seemed to agree. My thoughts immediately came to life, telling me I would have my little Chilean girl all to myself. She bade me wait by the alcove whilst she went to change. When she appeared, I opened the door in my politest manner; with a smile she walked past me, leading the way to the street entrance. As if I had ordered it up drew a taxi. I wondered if my little Chilean had ordered it, as I had wished for a nice romantic walk through the darkened streets. But living up to the night club life as if I were a film star, escorting my leading lady, I entered the taxi, taking my seat by her side, and the driver closed the door and jumped into the driving seat and drove off without any instructions from me, as if he knew where he was going.

'Señorita,' I murmured as we sat together and my arm stole round that dainty little figure, 'Will you marry me and I will take you far across the sea to London, that great big city,' and I took her finger pretending to put a ring on it, then pointing to us both. 'My sweet Señorita, your eyes are like the stars that flash in the night, your hair is of the softest silk, your loveliness will be unsurpassed in the whole of that great city. I will rise to be a great Admiral if you will be my Lady Hamilton.' She did not understand the words I spoke, but guessed something of their meaning and put a restraining hand on the lips of her amorous sailor, and bade me keep my seat to wait until our limousine stopped for a more favourable answer, her little hand squeezed mine.

I sat back patiently to wait. I had no idea where the taxi was going, nor did I take any notice as I was so wrapped up in my romantic bliss, but now it turned a corner rather sharply and drew up by some tall buildings. Immediately the driver asked for his fare, without getting out, which I proceeded to produce, fumbling a little in my hurry to pay, as my Señorita had now quietly opened the taxi door and was disappearing round the back of it. The driver seemed very slow to settle up, and wanted another note before being satisfied. I jumped out in a hurry just in time to see the legs of my lady love fast disappearing up some steps. Swiftly I followed, to enter a narrow passage where steps and passages led off either side, but not a glimpse or sign of her remained.

I walked back to the end of the passage to where I had left the taxi, but that had gone. I sat down on the steps cursing myself. For thirty minutes I remained there, hoping my lady might return but eventually my deeper feelings, which I knew were right but which I often ignored, told me I had lost her and it was no good sitting waiting on those steps any longer. Dejected and despondent I picked myself up at last to make my way through the narrow streets, with their high buildings.

I began to wonder where I was in this large city, and which way it was to the harbour, as there was a place close to the harbour where we had booked a room for the two nights of our leave. There was not a soul in sight. I walked on and looked up into the sky to see the stars twinkling down. Suddenly I remembered I could get my position from the Southern Cross, and set my course accordingly. A little bit of navigation in the heart of a city! At last I came to a spacious opening. I placed the Southern Cross over my right shoulder and set off in that direction. It must have broguht me some kind of luck, as I then came upon a shipmate who was also stranded. He told me he had been in the night club, but on leaving and getting into the open air he didn't know what hit him. His head was aching, and he asked me to find him somewhere to rest, I told him I was making my way south, keeping the Southern Cross over my right shoulder when it was possible to see it, and if he cared to put himself under my command and follow in line, we should reach the harbour before dawn. He looked at me oddly and said he thought the drink had made *him* bad enough, but I seemed completely gone. However, he would follow me if I could find him somewhere to rest his weary head.

On I went, with my mate following astern, when quite suddenly I came upon a lighted doorway with two policemen standing by. Just the place for a night's rest. Walking up to them I sat down by the door and showed them by laying my head down that I needed some sleep. 'Two of us', and I lifted two fingers as my mate came staggering up, in no fit state to carry on much further. I think they had a little conference between themselves, then one beckoned us to follow him. Over some rough stone floors, down into a deep cellar which seemed to be full of doors and iron bars, opening two doors he showed us two cells which had hard wooden benches as beds. Nodding our approval we pulled off our jumpers, rolled them up into a pillow and stretched out on the bench and were soon fast asleep.

Someone was trying to get some life into me as I came to, in stupefied condition, gathering my dazed senses together in a rather lifeless way. Suddenly I sat up with a jerk to find I was in a prison cell with a constable shaking me. It was a rather alarming predicament to find myself in. 'Had I done something wrong?' I quickly came to as I was now thoroughly alarmed and ran over in my mind the events of the previous night. Then I remembered how I had stumbled on the police station whilst guiding a shipmate through the streets. Slowly I rose to my feet, as I felt in a very poor condition. The constable was trying to arouse my chum with very little success. All I could hear were grunts, groans and curses. It was fortunate that at that moment the policeman could not understand the words tumbling from the mouth of my shipmate as he tossed his arms and turned over, wanting to be left alone to continue his stupefied slumber.

The constable turned to me, throwing out his arms and putting on a hopeless expression, as much as to say 'I give up'. He then picked out a wooden bucket, motioned for me to follow him which I did rather reluctantly as I had visions of having to scrub the cell out before getting my release, but these doubts were very quickly set at rest when we came into a courtyard with a hand pump, and he gave me to understand I could get a wash. Then I quickly dressed and returned finally to my chum to give him another shake, just to see if I could persuade him to do likewise, but he seemed to be past rousing. I left him to sleep on, then thanking the policemen for their kindliness in supplying me with accommodation for the night, I offered them a cigarette which they took gratefully, so I left them the whole packet. I said 'Cheerio' and walked away, leaving them sharing out the remainder of the cigarettes.

I found my way back near to the harbour where a large number of the boys were making a meeting place in a rather oversized room in one of the harbourside drinking places, which I gathered was made use of by most of the merchant seamen. There we talked over our night's adventures. I was reminded I looked terrible, as if I had been half submerged. After being reminded of my appearance a number of times, I finally walked out into the street and spotted a hairdressing saloon. I walked in and being the only one in the shop, the proprietor bade me be seated in his customer's chair. I pointed to my chin for a shave and for my hair to be closely cut, and with his nod of approval he set about his task. He gave me a shave and a haircut in a very short time. Then spreading out his hands, he asked me if he could do something more. To this I nodded my head and he quickly got working on my face with an electric massage, rubber sponge, and hot flannels. I sat back in his chair, really enjoying it, feeling better and more refreshed every minute. For fully half an hour he worked on my face and I finally rose from the chair feeling a different man. He charged very little, but I paid him handsomely and went back to my friends. They now changed their remarks to a more favourable impression of my looks.

I said to myself 'Half of the leave gone; I will finish the remainder in a leisurely manner. No more night clubs. I will make sure my room is waiting for me to claim it, then have a look in at the Sailor's Mission', about which I had heard the other boys talking. In most of the large trading seaports there is a small club run by Toc H or Sailor's Mission, where one is kept on the straight and narrow path by the kindly people who look after the sailor's welfare. Knowing the pitfalls and low dives that exist along the waterfront, they give you good advice on how to steer clear of these unhealthy spots and warn you of the consequences.

That morning as I stepped into this Sailor's Mission fresh from my grooming at the hairdresser's, and the effect of the night's celebration wearing off, I was offered a nice cup of tea and something to eat which would be prepared in a few minutes. 'Make yourself at home,' said the lady, handing me a cup of tea, 'I won't be many minutes before I have something for you to eat. Write a little note home, I'm sure your folks would be pleased to know what you are doing in this city so far away.' I thanked her for the tea and just nodded when she pointed to the writing paper and envelope as I felt a wee bit guilty about writing home and telling my folks what I had been doing.

'Well, well, here he is repenting at the Sailor's Mission after last night's sins' – my privacy was invaded by an inrush of my nightclub chums from the night before. 'I have no sins to repent, misdeeds yes,' I replied. 'Be seated and pay a little respect to the Mission before commencing on another wild night. The way you talk of the girl back home, and the way you behave out here, some of you have no principles. It's all right for you to drop in here now, but tonight where will you be again? In the Trocadero staring goggle-eyed at those Señoritas. Don't you understand they don't want you, only your money?'

We were all happy and enjoying our leave, and we chaffed each other with youthful enjoyment as we ate the cakes and drank tea, played cards and table tennis, passing the time away whilst we waited for the lights to come on, then we gradually drifted off in twos and threes towards the sparkle and gaiety.

At the night club which I had visited the previous night, an absorbing discussion was cut off abruptly, as two little hands were clasped over my eyes. As I removed them there stood my Señorita of the night before, all smiles, and tenderly nestling her little head against mine.

'Why did you leave me last night?' I tried to make her understand how annoyed I was, and disappointed at her sudden disappearance, but the more I tried, the more she kept smiling and petting me. I think she knew all right, but she pretended not to understand.

'Come on, sit down' – I ordered her a drink, 'You're just a butterfly, dazzling me with your pretty colours to make yourself so attractive that my eyes follow your every movement. You need someone to clip those pretty wings. If I had lived in Drake's days I would carry you off to the ship and make you wait on me and look after me like a slave then I would throw you off at the last port before sailing for home, but in this modern age I'm not allowed to do that. Thank goodness you don't know what I'm saying!'

Although we didn't succeed in our endeavour to make them think they were being entertained by the cream of the high-ranking officers of the Royal Navy, we did enjoy their company and got a kick out of it. I think our Señoritas valued our friendship and the fact that we respected them in our behaviour won their admiration so much that I think that had we been able to carry them off, they would not all have resisted.

The night wore on, all was merry and the place was alive, filled with the sound of voices talking and laughing in Spanish and English. We had decided beforehand that there would be no racing around all night, and that we would leave about midnight in order to get one night's rest and make use of those beds which had been booked, so very much to the surprise of our young ladies we bade them farewell and promised to see them again before the ship left. We strolled back to our harbourside hotel, turned in our beds – just like respectable citizens, as one of the group remarked. We all slept peacefully and were awoken by a pot of tea being brought to our room. We all arose with plenty of time to refresh ourselves, had a nice breakfast, a last drink together at the bar and then walked back to the ship, arriving just a few minutes before our leave expired.

CHAPTER NINE

We changed into working clothes again at the same time as the other watch were changing to go to their 48 hours' leave. Many questions were asked as to where to go, and what to do. We gave them the best advice we could. 'First, don't go to the nightclub called "The Trocadero", don't indulge in any intoxicating liquor, keep to the Sailor's Rest, book rooms and turn in promptly at ten each night.' 'We can always do that at home', was the answer as they dashed off to be mustered and inspected before commencing their leave.

When we were talking over our experiences some days later, I remarked: 'I've only done two years of my seven as yet.' 'Two years,' my pal remarked, 'Do you regret joining up?'

'No I don't,' I replied. 'I very often think of home and it makes me wonder how I happen to be all these thousands of miles away and seeing all these wonderful parts of the world in such a short time. If you had mentioned the names of these countries that we have visited to me, before joining up, I wouldn't have known where to find them in the atlas, and as for the Navy, I'm the very first of the family to join. Why, I don't know.' I sat up and threw my head back with a proud smile on my face thinking I had done something great in breaking the family tradition, and in being accepted in the Senior Service.

'You see, all my family have been in the Army, and living so far inland and born to the soil, as you might say, it rather gives me a kick to think I have lifted the status of the family up a peg.'

Further discussion was cut off as a messenger dashed up calling out my name. 'Yes, will you report to your Divisional Officer with Ordinary Seaman Dowding. Right away. You will find Ordinary Seaman Dowding waiting for you amidships.' 'Now what can that mean?' All the things you have done wrong immediately spring into your mind. I couldn't think of anything with which I could be charged, so I jumped up smartly, making my way to join my companion who, like myself, was at a loss to understand why we had been summoned.

'Well, we will go and face up to it,' I said, standing before my officer's cabin as I gingerly tapped on the door. It was thrust open and I announced our names. 'Told to report to you, Sir.'

'You may come in,' replied our Divisional Officer.

'Thank you, Sir.' Inside, seated at his desk, with his attention fixed on a sheet of paper with the P.T. Instructor bending over his shoulder after opening the door, they both studied the list of names on the paper. We stood silently by as between the two they were selecting names, and saying to each other 'That will complete the team'. They turned to us.

'Sorry to keep you waiting. The P.T.I. and myself have carefully given our consideration to the best cricket team we can select from the whole of the ship's company. It is a very important game as we are going to represent England against Chile in an all-day match to be played on their best ground, which has a large stand, and we hope it will be watched by a good-sized crowd. We know there will be several hundred English people as cricket is played regularly, and some good players with professional experience will be taking the field against us. We have both taken notice of your playing during the inter-port games, and have decided that you two young players will be included in the team. I know both of you are keen and will give of your best, but in this game your best is not going to be good enough. We want that extra bit of something which makes a good player into a star. Both of you have got to bring that out on the day of the match, and to do that you must be fit as well as keen, so if you can stop on board for a few nights – no runs ashore until the match is over. We don't want to stop anyone's leave, we have more faith in you than to do that, but just a word of warning. "Wine, Women and Song" and an important match the next day just don't work out. Besides the Captain and all senior Officers there will also be watching the Mayor, the Ambassador and all the important officials of this town. A lunch will be provided which will be attended by all these, and you will be required to look your smartest, spotless in your whites. Think carefully and give me your answer now if you are willing to accept and co-operate to the best of your ability.'

I gasped out 'Yes Sir'. I think that I shook a little with

excitement before I overcame my feelings and took command of myself to turn to my other partner who had not, as yet, replied and stood solidly without showing any emotion such as I had expressed. I wondered for a moment if I was wrong, and why I couldn't be as steady as he was. A shadow of doubt crossed my mind – was he going to refuse? Perhaps it took longer to realise the significance, or was he just shy? The Officer was looking at him saying 'Think it over carefully. Take your time.' At last, when he said the all-important 'Yes, Sir', he said it with a very firm voice and I knew he was determined to do well.

'Thank you both. Anything further you wish to know, or if you want any help, get in touch with the P.T.I. or myself. Tea and the time of leaving will be posted on the notice board.'

We tumbled out of the cabin with a polite 'Thank you Sir'. I rushed off to the mess full of the exciting news, but then I pulled myself up and thought it best to be like my chum and take it calmly, so I said very little about it. One or two of my messmates did remark that they had seen my name down to play cricket and raised their eyebrows, which was more polite than saying how surprised they were, and a few times I was referred to as 'Jack Hobbs' (who had been England's star batsman).

As the next few days slipped by I faithfully stayed on board and was all ready on the morning of the match when the P.T.I. took charge of his representative team of English players and we boldly walked over the gangway, led by our gallant Instructor to a chorus of 'Quack Quacks' from the not-too-busy hands of the ship who somehow had left their work to gather along the ship's side to give us this far from encouraging send-off.

We entered a large stadium through a decorative gateway. It had seating for many thousands and a restaurant, bar and dressing-room, and all that goes to make up a first-class sports arena. In the centre was the grass of the cricket field, which almost equalled the green colour of an English county cricket ground. We were made welcome and shown to our dressing room, and the restaurant was pointed out where we could mingle amongst our opponents and their friends. Iced drinks with refreshments would be served at our request. We could inspect the pitch and make ourselves thoroughly at home as the match would begin at 10.30 a.m. and play continued to 12.30 p.m. then luncheon, starting again at 2.30 p.m. until the conclusion of the match.

With the rest of the team I sampled their refreshments, having a word of welcome with some of our hosts until I was informed that it was time to change and take the field. I felt a little overawed and kept close to my young friend. As soon as we were all ready, led by our Captain, we walked out on to the field to the accompaniment of the Marine Band who played 'See the conquering heroes come.' I felt far from being a hero and was glad it was some distance to the onlookers, at the same time I did feel a thrill, especially when there was a short ripple of applause, but my chum said 'It is not for us, it is for the band.'

We opened out, throwing the ball to each other as the two captains tossed to see which side batted and we were quickly informed by our Captain to remain on the field, the other team withdrawing to pad up for batting.

Lunch was a grand affair with all the nobility sitting at the main table. I found I was seated on the end of the third row of tables, a respectable distance from the centre of attraction, which suited me. After the excellent lunch we had a friendly chat, a game of billiards before going out on the field again to finish off our opponent's innings, and to walk off the field to a round of applause. Now it was our turn to handle that shapely piece of willow (which had written on the blade 'Made in England' and was autographed by Jack Hobbs). As I shaped up in the dressing room I remarked how good it was to feel my hands round a piece of old England again, and with it I declared I would defend the honour of the old country, wielding it like a knight swinging his sword in the olden days. 'You'll get your chance,' chimed in the P.T.I. 'You're in third wicket down. To get your eye in start swinging your bat to obliterate some of these mosquitoes.'

Our opening batsman was now facing the umpire, and at last acknowledged that all was ready and the bowler could commence his run. Our eyes were now on the bowler, standing such a long way from the wicket that we knew he must be fast. Now with a hop, skip and a prolonged jump he hurled the ball with the doubled-up fury of a tornado. The sudden onrush so took our batsman by surprise that I don't think he knew whether it was the ball or the bowler or both descending on him. All he could do in his surprise was to lift his bat a fraction from the ground, which was just enough for the ball somehow to find its way underneath and shatter the wicket in all directions. The bails flew high over the head of the wicket-keeper and our poor batsman with drooping head began his lone trek back to the dismayed and silent members on the balcony who had given him that rousing send-off a few moments before.

Our attention was now taken up by the outgoing batsman who had passed the incoming one by the edge of the field, and had cheekily asked him if the ball was keeping low. We had only given him a half-hearted clap as the blow just dealt to us by the quick dismissal had taken some of the sting from our clap, and a slightly creased brow had appeared on the P.T.I.'s forehead, and even the sports officer had left his important guest to converse with him, and I heard them mention my name, saying 'Do you think he's good enough to go into so high a position?' I held my breath as I strained to catch the answer, all my hopes seemed to be dangling on a spider's thread. I heard the other remark: 'If this one gets a few runs we will leave the team as put down, as it might unsettle the lads to change them round.' Just then the bowler was doubling himself up in all his fury to hurl his second delivery, and I prayed that something would happen to deflect it from that wicket. All was well. It went flying high over the batsman and wicket-keeper, right to the boundary for four byes. As quickly as our hearts had sunk with doubt, now they soared with renewed hope as we saw the score go up on the board, four runs. We must get a few more for me to retain my batting position. Steadily and very gingerly a few more runs came along but very soon, as if expected, down went another wicket by that demon bowler, and my mate nudged me to pad up in preparation to wield that willow, not like a knight swinging his sword, but in a steady hand to stem the onslaught. He remarked 'Remember in this match England expects you to do your duty, not only in the cabaret at night but on this blinking cricket field. Just show them.'

'I don't want to go back to the ship just to be called a flop,' I said as I trembled, and my voice seemed to flutter as the sports officer came up with the P.T.I. and said 'You've started to pad up. Stop a minute.' I thought, 'Now they are going to put me down lower.' Then I heard the P.T.I. say, 'You haven't got time now Sir, you will have to let him go in. The other fellow is already on his way back to the pavilion.' Then followed a scramble to get me fitted up, with the sports officer giving me instructions to keep my end up in a hurried and worried voice. When at last I stood ready, my only little bit of encouragement came from my young friend, who whispered as I walked off 'Don't take any notice, just play your game.'

The next moment I was standing on the pitch. I thought 'Damn the lot of them'. I don't know if anyone clapped. I don't think they did. I was angry and I think that helped me overcome the nervousness that otherwise I would have felt as now I stood my ground.

I saw a form come leaping towards me, a ball of leather flying down the pitch. I moved my bat in front of it, and it stopped dead. I turned and stepped back as if I had no trouble at all to face this bowler, and to let the wicket-keeper come forward to scoop the ball back again. This I did for the remainder of the over and two or three more overs were dealt with in similar manner until my anger began to vanish and I began to settle down and realise I could wear this demon bowler down. Doggedly and determinedly I stuck to my post, with runs coming slowly, with the occasional wicket falling, but now I had at the other end a partner who was quite the reverse to myself. Every ball that came his way he seemed to be able to hit, and now the game was beginning to liven up. Runs were coming fast, and even I was coming out of my shell and adding to the score. Our opposing skipper switched his bowlers, changed his field and did everything in the cricketing book to break our stand, even to bowling off the wicket for a catch, but to no avail. Gradually the score mounted until it was level, then passed and I had the honour with my partner of walking back to the pavilion being loudly clapped and cheered. It is a nice sensation to be so honoured, it is the moment you have dreamed of and longed for, to be the star and hero of a game, when your team mates slap you on the back and all around are full of praise. The sports officer came up and congratulated me. The P.T.I. said 'Well done. You saved the day. We are waiting for you in the bar where a bottle is going to be opened.'

There, with our opponents, we drank to each other in the most friendly way, and I felt sure that much good feeling would be carried on between the two countries for some considerable time, and trade would benefit, especially in Scotch Whisky. Thank goodness those taxis arrived to carry us back as with many a hearty slap on the back and a good many long lingering handshakes we were shepherded inside our transport by the now anxious P.T.I., who was responsible for our return to the ship. As he closed the door one side, one of those Scotsmen would be round the other side pouring out a wee drop, one for the road as we drank and reminded them to get back to bonnie Scotland, may the heather grow upon the hillside. Scotland forever, and by yon bonnie banks and by yon bonnie braes. At last the P.T.I. stood on the running board and made each driver get going before jumping off to get the next under way. As we drew nearer to the ship we said to each other 'Pull yourself together, keep steady and don't forget to salute as we step on board.'

The taxi I remember came to a stop close to the gangway. Very carefully I stepped out to hold on to the hand-rail. Then step by step holding myself erect I walked up that gangway, standing and saluting as my foot touched the ship. The Officer of the Day had walked over to the other side as if he didn't want to welcome his victorious team home. I turned and began to walk aft until someone gently took me by the shoulders, turning me round saying 'You haven't got any rings on your arm yet, chum.' I thanked him and said, 'We won the match. Where are those who said "Quack quack" when we left?'

'Right for'ard in your mess.' Two of my messmates on either side led me to my mess, making me comfortable so that I could sleep off the effect of my celebrations.

Goodbye to Valparaiso, out to sea again steaming steadily back to Bermuda. Shaking off the effect of our ten days stay. A large number of people had lined the jetty waving us farewell, including some of our cabaret girls. One had presented a note to the quartermaster on the gangway. It had my name and the number of my mess written in lipstick. One of the messengers had been sent to deliver it, saying 'A lady brought this and is waiting by the gangway to see you', putting the note in my hand.

'Wait one moment,' said I, looking at the note and now remembering my frolic ashore. Hastily I drew a heart pierced by an arrow and wrote 'Farewell, farewell, I am too heartbroken to see you.' 'Please give her this and make her understand by some means that if I saw her, I would have to jump ship as the parting would be unbearable.' 'Right-ho, but I shall want a drink of your tot.' 'Right, you shall have anything if you can just pacify her. I couldn't have her crying on my shoulder in front of the whole ship. They might shout out "Marry the girl".'

He looked at me oddly. 'I hope you are an honourable sailor.' 'Of course I am,' I stammered. 'She is the Señorita from the cabaret.' 'Well, I will do my best, but I think you ought to face up to your responsibilities.'

He returned five minutes afterwards, saying, 'Your Señorita

won't leave the gangway, and keeps repeating "Captain", so the Officer of the Day wants a word with you to see what this is all about.'

A bit alarmed, I followed him to the quarterdeck to be interviewed by the Lieutenant who produced the message saying,

'This is your name and writing?'

'Yes Sir.'

'Why does this lady want to see you and keep on repeating "Captain"?'

'It's like this Sir, I went to the cabaret and had a few drinks and danced with her and wrote my name and mess on that note.'

'You're sure that was all?'

'Yes Sir.'

'Why does she keep repeating "Captain"?'

I hung my head and moved my feet and tried to utter something.

'Come on, out with it. There must be some explanation.'

After a pause I did get out, 'Well, Sir, it is like this, after a few drinks one begins to say things which one wouldn't normally, and I remember saying something that I was next to the Captain if she came down to the ship. I didn't think she understood what I was saying.'

'In future have less to drink on your runs ashore, then your tongue won't waggle so much and get you into difficult situations.'

'Very good, Sir'; I hardly knew what to say.

'You only have five minutes before we sail, just go and try to explain to this girl friend of yours, and take this.' He thrust that paper in my hand again. I saluted and doubled smartly away to stop by the gangway. She stood on the jetty looking across. Spotting me she made as if to come on board. Hastily I ran across myself.

'Ship leaving,' I cried. 'Going far out to sea. No Señorita on board. Terrible sailors, very bad men. Take this.' I gave her back the note pointing to the paper saying 'Ship return. I will come and see you.' She looked at me and pointed to the ship, then to herself. I shook my head saying 'No' at the same time. 'I must leave now, I will see you when I come back to Chile. Goodbye,' and I swiftly withdrew over the gangway again. Fortunately everybody was busy with their duties, and with so many others waving 'Goodbye' it went almost unnoticed except for the messenger, who promptly came round for a drink of my tot.

I now sat on the deck watching the fading coastline, getting smaller, then disappearing altogether to leave us all alone in that vast ocean again. I looked down into the sea to watch it sweeping by as the ship steamed steadily on. The routine of the ship was occupying our time and minds. I slung my hammock, lowering myself in and let the steady roll of the ship soothe me to sleep.

CHAPTER TEN

A few days later, we learned from the information on the daily notice board that a small island called Juan Fernandez was due to appear on our starboard bow about noon. This was the actual island where Robinson Crusoe had been shipwrecked.

That morning no one moved to follow when our P.O. went below for his grog. We were altogether scanning the far distant horizon for the first sight of the island. We had all stopped work and one of the Petty Officers was telling us of his far-off days when, only a boy, he had served in one of the ships which had taken part in the Battle of the Falkland Islands. After inflicting heavy losses on the enemy one ship had escaped. The fastest German cruiser called the *Dresden* was hunted by our ships, and was finally caught up with at this small island. As this ship rounded the island and spotted the mast of the *Dresden* they had steamed straight up the bay for the final kill, only to see the white flag being hoisted, gaining time to blow up the ship which now lay a half-submerged wreck. We would see for ourselves as we went in.

At first we had gathered round listening to while away the time, but as the story became more interesting, our imagination was caught as the P.O. became absorbed in his tale of boyhood service in wartime days, when the navy really went into action, when flames belched from the guns as they fired in anger and the enemy were firing back with shells bursting all around. 'Not like you youngsters are doing now, just closing up doing dummy runs at imaginary targets. Firing a salvo once in a while when half of you flinch and duck your heads at the sound of a bang.' His eager young audience now listed with silent attention, taking in every word. Eager questions were fired. 'What was it like to be in real action?' 'Did you ever duck your head, P.O. when the shells whined over?' 'Do you think there will ever be another war?' All of us were asking questions as he stopped talking, eyed us for a second and paused to let his words sink in, noticing that we were all burning with zealous attention with work and everything else around us forgotten. 'War?', he continued, 'Lord help us. In those days we had wooden ships and iron men. A few of us old ones remain and then I fear for England.' He seemed to eye us with a doubtful look, just as if we would never reach the standard of bygone days.

From the bridge we heard the lookout report: 'Land on the starboard bow, Sir.' We all looked at our P.O., who was standing perfectly still with his weather eye searching the distant horizon. A minute or two passed, then one of us spotted it. 'There it is, P.O. Your island which you shared with poor old Robinson Crusoe.' Slowly we all moved towards him as now a distant blur was beginning to take the shape of a mountain as we drew nearer. We could see two high ridges, then the valley in between covered with dense foliage. 'That's how I remember it.' Our P.O. began to address us. 'In another few minutes you can make out a smaller ridge, then the bay, a clear space, then a few houses way back or thatched places of some kind if I remember correctly.'

'Probably the remains of Crusoe's hut,' was the response. 'Ah, there it is,' was the heartening cry of keyed-up delight, bursting forth from our most solemn P.O. as he pointed to some object which was just visible. 'The wreck of the *Dresden*. Still there after twenty years. Gather round my hearties, there is proof of what took place when last I sailed into this bay with the white ensign flying at the mast, head guns loaded and at the ready. Just the sight of us and the enemy scuttled itself.'

There before us lay the wreck; a twisted rusty iron hulk was all that was left of the once proud ship that was the pride of the German Navy. Her speed had enabled her to leave all the others behind, but lack of fuel had been the cause of her doom. Slowly we approached the heap of old iron, its jagged and burst plates now visible to our naked eye. We all stood still and silent as if this was a monument and we were paying our respects as the *York* now reduced to a few knots passed by; and there on shore was a large monumental stone draped with the cable and anchor of the ship, and hanging over the arms of the anchor was one of her lifebuoys with the letters reading 'S.M.S. *Dresden*'.

A confident P.O. now turned to us with his chest and shoulders expanded to their full capacity, his voice had a triumphant ring as he stood erect and ordered us to prepare to anchor, remarking 'You've seen enough for one day of what happened to our enemy when ships were manned by men, Get moving, Stand by that cable, you've got something to do if you ever make the grade.'

When we had anchored and all had quietened down with the Captain leaving the bridge, I remained standing on the foc's'le drinking in the beauty of the silence and peacefulness that seemed to hang over this little speck of earth. Our presence had not even stirred the inhabitants. I could only see two figures moving on shore whilst further back a man riding a horse was leaving a building, making his way towards us, whilst up on the mountainside several sheep or goats stopped stretching out their necks as if to take a sniff of the air, or a look to see what monster had arrived to upset their tranquillity. I thought 'What a paradise it would be to live on this island, and what a story I will have in years to come if one day I have a grandson.'

I smiled at myself at my thoughts. I was only twenty-one, too young to let such silly thoughts enter my head. Slowly I straightened myself, left my daydreams and joined my messmates to be told there was recreational leave of up to two hours. About fifty of us gathered by the gangway; we were told that a black flag would be hoisted from the yardarm to recall us, and if anyone didn't see it and return promptly they would get left behind to fend for themselves like Robinson Crusoe, and only one ship a year put into this bay.

Now the ship's two motorboats began plying to and fro, filled with libertymen. As a junior rating I was in one of the last boat loads to leave. We drew up alongside a large wooden jetty which looked as if it had been made of tree trunks cut from the island, as some looked very fresh. I stepped from the boat on to this tree-trunk platform to walk its length of about twenty yards, then I stepped on the sandy shore to feel my feet sinking in the loose sand at each step I took.

Leaving the party I turned, walking along to the nearest part of the mountain. If I could make the summit of the first small ridge, which didn't look very high to me, I could stand like Crusoe did. Perhaps this was just where he used to walk every day to watch for a ship. A half hour at most should see me to the top. I reached the trees and shrubbery, beginning to climb. It was quite easy. The goats had made it clear in patches, so by zig-zagging round big boulders, following the trail of animals, feeling full of energy, mountain climbing seemed nothing. I turned to look back and there was my trail, leading back along the shore, each step was deeply impressed in the loose sand. 'Not much chance of getting lost,' I thought, 'leaving a trail like that.' Onward, up to the top. I startled a goat on rounding the next big rock and noticed it had long horns. It sprang away for a few yards, then turned and looked at me, lowering its head, and for a moment I thought it was going to charge. Then it turned away and strolled off, much to my relief. I didn't want to be butted by a goat with horns a foot long. Besides, it might be waiting in those shrubs further up. What would Crusoe do in this situation? Luckily I had my sailor's knife, so I hacked away at a large bough until it snapped from the tree. I trimmed the leaves off and armed myself with this, then advanced once more. In places it was getting quite steep, so I pulled on the bushes to help myself up. Stones rattled down from under my feet. I began to think 'I've been climbing half an hour and I'm not half way up this small part of this mountain. Looking from the shore I gave myself about ten minutes to reach the top. I'm not going to let this beat me, but two hours will soon pass.' I spurred myself on, walking along the goat trail, pulling myself up as convenient holds presented themselves. Another ten minutes and I was making progress, but not so much as I had hoped for. I had now come upon a thicker patch of foliage, so I had more hold and was getting on quite well when suddenly a bleating noise sounded right behind me. I jumped round clutching my cudgel, falling sideways as I lost my hold. Two sheep went rushing past. I had disturbed them as they had been lying in the shade in this thick patch.

I picked myself up. My arm was bleeding from a graze where I fell on some stones, my trousers were torn and I still had not reached the top. I bent down, eyeing my torn trousers, when with a fluttery swoop, followed by a din of screeching, a flock of birds flew into a tree not a dozen yards away.

What with the goats, sheep and birds, and being all alone, I decided it would be better for me to abandon my enterprising adventure, besides there might be some larger animals, perhaps much more fierce, or snakes, which hadn't crossed my mind until right now, might be hiding in the thick tufts of grass. With all these doubts and fears now entering my mind I began to think I was silly to be all on my own, and what about the wild hog?

According to the book they did actually live on this island. They could be fierce, that decided me; I retraced my steps, came in sight of the bay where the ship was still lying peacefully at anchor, no flag was flying to recall us, and down in the valley several figures were strolling around as if not knowing what to do with themselves. On seeing this reassuring sight from the mountainside, I felt happier, and sat down to rest. I made my way down to the spot where I had commenced my climb. A few ratings hung about waiting for the boat to take them back on board. Apparently their walk had not been very eventful, the only thing of note being the seamen's memorial. 'Where is it?' I enquired. 'Just up that green grass track – not far if you want to see it.' Off I set to come upon a little group gathered round a large stone; embedded in the stone was a tablet with this inscription on it.

In Memory of
ALEXANDER SELKIRK
A native of Largo in the county of Fife, Scotland
Who lived on this island in complete solitude for four years and four months.

He was landed from the Cinque Ports Galley 97 tons 16 guns A.D. 1704 and was taken off in the "Duke Privoteer" 27 Feb. 1708.

He died Lieutenant of H.M.S. Weymouth A.D. 1728 aged 47 years.

This tablet is erected near Selkirk's Look-out by Commander Lowell and the Officers of H.M.S. Topaz.

This man was the real Robinson Crusoe. From his experience the story had been written. I tried to imagine how he had lived, getting food and water and keeping a lookout. This must have kept him occupied for the greater part of the day. After those years and months finally sighting a sail, and to see it heading for the bay. To hear once again a human voice, he either went mad with delight or knelt down and thanked God for deliverance. I thought of this man's endurance, then of my own tame affair of the past two hours. No wonder the P.O. shakes his head when we fall in for work; perhaps there is something in what he says, and we are not so hardy as those who sailed before the mast.

Now someone shouts that our recall flag is flying and the motor boats are ready to carry us back. On to the landing stage, stepping in the boat, I say farewell to Juan Fernandez. I am pleased that I have been to this little spot. It is not always the big names and places that run the world, these little isolated spots play their part in the world's history.

Sailing on back through the Panama Canal to our base at Bermuda to prepare for our next adventurous cruise, I am training hard for the sports. I do well at running and am included in the last fifty to represent the ship. During our last voyage I have been asked if I would like to go for a higher gunnery rate. I reply 'Yes', but think no more of it. I bathe in the clear blue sea, play games and do my running exercises every evening. The shore canteen I visit as my pocket will allow. I feel fit and strong, keeping myself in tiptop condition. There are no cabarets or females to distract Jack's attention. The capital, Hamilton, is, I understand, a high-class sea resort for wealthy Americans. As I am most likely to be out here for another two years, there is plenty of time left for me to have a look around later.

I return one evening full of good spirits, having beaten two runners who had been twenty places in front of me in their placing in the ship's team. I go to the bathroom for a shower where I am told the Chief Gunner's Mate has been looking for me, and I am to report to him immediately on my return. 'Whatever can he want?', I think to myself, 'some change of Gunnery Station?' I presented myself at the wide-open door of the office.

'I have been told to report immediately Chief.'

'You have to be ready by 10.00 hours on Wednesday morning with bag and hammock to take passage to U.K. There are ten others besides yourself going.'

'Why me?', I asked.

'It is for a higher Gunnery rate. You will be doing a three months' course for a Seaman Gunner. You have just a bare arm. Don't you want to see some badges on it? Not many barmaids in Chatham pubs will look at you twice if you haven't got something on that arm.'

I turned away. Did I want to go home or remain? I didn't know. I thought of home and a happy feeling spread over me. But wouldn't it be best to do my two and a half years' commission with my shipmates? The ship was going to visit America and Canada. There was time to put in a request to ask to remain. Seriously I thought it out but did nothing. Fate must take its course and I soon found myself saying 'Cheerio' to all, shaking hands with Shorty as I told him perhaps next time I saw him he would be grown into a full-sized man.

CHAPTER ELEVEN

We went down the gangway, carrying our bags shoulder high, piling them on to a handcart, returning for our hammocks and small cases. When all was loaded we pulled and hauled to get it going along the jetty; the quiet rumble of the wheel was the only noise, as no one spoke. I turned and looked at the ship I had just left. I ws saying 'Goodbye' to my first ship.

A fresh ship awaited me for passage home to England, and from the regret at leaving one ship, a thrill of pleasure now ran through me as I carried my bag and hammock aboard. Quickly we were dealt with, each given a mess and part of the ship to work in according to our respective branches. The ship's company had completed their commission of 2½ years and were thinking of home. Little attention was given to us for in a very few hours it would be under way on its last trip. It was a West Country ship, whose home port was Plymouth. We were told to keep out of the way until the ship was at sea. This we did quite successfully for three days, apart from going for our meals, giving a little help in the mess until we were rounded up to make sure, I think, that we hadn't got lost overboard. We were then given a little cleaning to do, the small amount of work we had to do making the day seem to pass very slowly.

Soon only three days remained before we were due in at Portsmouth. The ship stopped there for a few hours and we were told to be ready to disembark, to take the train from there to Chatham, our own home port. Two days to go, the flush of excitement began to spread over each member of our little company, one day – and our step seemed lighter. We trod on air instead of the deck, meals were no longer necessary, we didn't know whether we ate them or not. Then the final day arrived at last. We arose from our hammocks – sleep not having closed our eyes all night – and a hum of delighted murmuring was heard from our little group long before bugle call sounded.

I was washed and shaved and passing up and down amidships long before the light of dawn broke in the eastern sky for England lay before me, just a few hours away. Up and down, about turn, long minutes of endless pacing, but at last a faint streak of mist seemed to break through the dark night. I could see merchant vessels of every nationality. Our merchant vessels dipped their red ensigns in salute to the white ensign of the Royal Navy – a friendly gesture of respect towards each other who both fight and know the perils of the deep.

The loudspeaker was calling our attention, eagerly we turned our heads to catch each word. 'All taking passage be ready to disembark in an hour's time, when the ship will put in at Portsmouth. No leave will be given to the ship's company as the ship will put to sea again at once, sailing for Plymouth.'

There was no need for us to get ready – we had been so before the break of day. Only another hour, down to the mess to say 'farewell'. The whole messdeck was filled with all hands waiting to line up to enter harbour. An order was given, followed by the mad scamper of many feet hastening to their stations.

We remained on the messdeck listening to the command being given, the prolonged stillness as salutes were exchanged, then the sound of running feet, telling us at last the ship was being secured. As all portholes are closed and those only in the rig of the day in perfect station order are allowed on the upper deck on entering or leaving harbour, we patiently bided our time eager to see an English seaport again. No longer can we remain hidden, so cautiously we ventured out from our concealment to see a skyline of masts and funnels. Now we lined the guard rail looking at our own naval port with the experienced eye of one year's foreign service to our credit. We hardly had time to look round before our next order. 'Tug on port side. All taking passage report amidships with bag and hammock.' There was a scramble to be the first as we grabbed our gear and raced with it to the ship's side, handing it to the willing hands, lining up when finished as the senior member was given our papers and vouchers for our train journey and entry into Chatham barracks. We went down the ladder into the tug as it circled and then eased to a stop. Then it was up the stone steps, with my feet at last resting on English soil.

On the train, you open the paper and see the familiar headline print, 'Too much is being spent on our Forces'; 'Churchill the warmonger'; 'bread and butter before guns and ammunition'. I began to read and take interest in the world news, for now I have travelled and my experience has given me a small insight into foreign countries of which the majority of people in the country only learn what the papers want to tell them. I feel that I have got somewhere and become something, and that I have made some advance. So happy and contented with the swaying of the rocking train I fall asleep.

The next thing I know is a clatter and bang, followed by a big jolt. Then someone's hand is on my shoulder. 'Wake up, this is London. No time to sleep. Gather your gear.' We are transported from one station to another and take our seats in the electric train. The small stations rush past – remembered names are read as we flash by. A glimpse is caught of the muddy Medway as the train rumbles over the Rochester Bridge, stopping at Rochester a few minutes. We stand and get together our coats and cases, and hear at last the old cry of 'Chatham'. The door is pushed open with a cry of 'Hurray'. We jump on to the platform, and who do we see but a number of new ratings of a few months' service standing on the platform in their naval-made uniforms, hats on straight with the 'M' of *Pembroke* straight over the nose betraying their recent entry. Haughtily we swagger by, tilting back our hats, our muscular, bronzed athletic figures rolling with the gait that comes from so much sea time, feeling very superior and looking down on our fellow men.

Our baggage is soon pitched in the waiting van and we are off past the old Chatham Town Hall, down the dockyard road, entering the main gates of Chatham Naval barracks H.M.S. *Pembroke*. We jump down to muster by the guard house to do our joining and foreign service leave routine. We rush from office to office so that we can start our leave. I hardly have time to look around. We are victualled in Anson Block – one higher than in my new entry days when I was just an ordinary seaman. At last we are told that we are to muster at 10.00 hours tomorrow at the Leave Office in the drill shed to pick up our pay and travelling vouchers. We can have the ordinary night's leave up to 07.00 in the morning, and I decide to have a quiet run ashore to see the inside of an English pub and to get accustomed to the old ways. It would have to be quiet as I am almost broke. Tomorrow will be quite different, with our victualling allowance and leave pay we shall feel like Admirals. Preparing to go ashore there is quite a lot to do and I must order a new suit. I walk out, presenting my card at the main gate – no mustering at certain times and being marched with arms swinging down to the main gate as we had to do in our New Entry days. The night leave starts at 16.30 hours with a rush of many hundreds of married locals who have permanent homes or who have taken residence for their period of stay in barracks. A large number live in surrounding towns and it will be quite easy to visit London by train.

Joining the mad rush of liberty men I go through the main gate to the overcrowded dockyard bus with standing room only. I alight and walk into my special naval tailor's shop, 'Greenburgh's', where I have a monthly allotment to keep me in the best up-to-date, cut-to-measure, nipped-in-waist jumper, tight-hipped trousers with large bell bottoms to match and collars of the lightest blue. Many times I had been asked by one of the fair sex how I got my collar such a nice light blue and I would answer that it had paled by the hot African sun. The hats too were shaped and streamlined, deep enough to fit, not like the Purser's type which were round and flat, shoes were light and almost pointed, not rounded which we were supposed to wear. The tape measure is run over me many times to get my exact measurements as in some places I have bulged out and in others I have tapered in. Much concern is shown in their work to put me at my ease, making me feel that they are as much interested in turning me into a streamlined sailor as I am to be one.

I felt in a satisfied mood as I walked down the street of this famous seaport town where many hundreds of sailors like myself had returned from Foreign Service and were overjoyed to be back. Tomorrow I would be on my way home. I booked my room, put down for a call, and walked out. I passed the old theatre where the mule had almost been subdued by the juicy carrot, along to the pub where we had our farewell party. I entered the swinging doors, paused and looked around, not much doing tonight. The barmaid was just leaning on the counter at present. I walked across and ordered a pint of bitter. Picking up a glass mug she drew it off with a well-practised pull on the handle of the pump, making a large, frothy head, at the same time raising her eyes to give me the once-over. 'Hello,' she said, handing me the pint. 'Hello,' I replied and held it up to the light. 'It's grand to see a good pint of English ale again so clear and sparkling. It's the first one for over a year.'

I had a long friendly chat with the barmaid who finally

astonished me by saying: 'Promise me after this drink you will turn in for the night.'

I was taken aback by her concern for my welfare. It was kind and well meant. I wasn't a bad judge of character, but it touched my dignity as I replied: 'Don't you think I can take care of myself; do you think I shall fall to the first temptress to give me the eye?'

'Don't get me wrong. I want to help, but drink can stir you into doing something on impulse that you would ordinarily never dream of doing,' she replied. 'Besides, you seem to have a lot of money and you can lose the whole lot in no time at all. Tomorrow you could be penniless and that wouldn't be very nice for you having to go home and sponge on your people.'

She brought a smile from me and I said to her: 'If you are worried about me falling for the first girl I meet and parting with all my money I can put your mind at rest, because all I have on me is the large amount of 1s 2½d, the change from my last ten shillings.'

She smiled. 'Thank goodness. Only I've seen so many make fools of themselves in their first night's leave, treating everyone and being carried off for the night to appear in the morning sorry and almost broke, before they have even set foot on a train. You do understand, don't you?'

'I think I do, and as the situation is like this I will pocket the small amount of money I have left and take your advice and go and turn in for the night.'

Her attention was now being rapidly taken up by other customers arriving, and this put a stop to any further conversation. I stood leaning on the bar, hat well back, as if observing the whole room, watching a typical Service tavern beginning to fill with soldiers, sailors and civilians, some greeting each other by their Christian names as if this was their regular nightly visit. Some had pints, others short drinks. A fair sprinkling of ladies were slowly mingling amongst the men.

I continued leaning on the counter with a drink now and again from my emptying mug and meditating, for I was really back in the life of old England, and there were the ladies the kind barmaid had warned me about. I said to myself 'Here's one they won't charm'. Hardly had this thought passed through my head when one of them rose from her chair, crossing over close to me as she put her glass on the counter to be filled, her arm touching mine as if by accident.

'Sorry Jack.'

'Oh, that's all right.' I made room. She turned.

'Haven't we met before somewhere?'

'I don't think so.'

'You remind me of a boyfriend I once knew. I think he was on your ship, H.M.S. *York*.'

She straightened up as if to get a better look at my cap ribbon, her face coming close to mine, allowing her fair hair to lightly sweep over my face,

'Is your ship in port?'

'No. It's thousands of miles away.'

'How come you to be here then? Are you on leave?'

'Yes, I am, but only for tonight.'

She now gripped my arm.

'You don't mind me leaning on you for a moment? I have something in my shoe.'

'Go ahead,' I said, as she tilted her leg up, removing her shoe with one hand as she gripped my arm tight with the other. She gave her shoe a little shake and replaced it.

'Thank you,' she said. 'I need someone like you to lean on. I was all alone. Won't you come and join me?'

'It is nice of you to ask a lonely sailor like myself,' I smiled, 'but in the circumstances I find myself at present, I couldn't even ask you to have a drink or pay for my own. I'm flat broke. That's speaking the truth.'

'Haven't you been paid yet?' and she let go of my arm.

'Not until tomorrow,' I said, 'without stopping in this town I shall make straight for the station.'

She picked her glass off the counter, turning to go back to her seat. She lingered a moment beside me to say 'Well, Jack, if you care to change your mind tomorrow you can always find me around.'

I picked up my glass and drained it, and glanced along the bar to see my barmaid. I guessed she had been watching me whilst she was busy. I held my hand up as if to say I had come through the ordeal intact, and she gave a little smile and wave in reply, as I turned on my heel and left.

After the train and the local bus, I walked the last mile home. A few cars passed by, a loaded lorry thundered past. I heard the drone of a tractor in a distant field. I thought 'Where are the horses I used to work with on the field? Has the engine so quickly ousted them from their job?' I rounded the bend to see before me the cottage. Its gabled upper window peeped over the tops of fruit trees in the garden. A yellowhammer swooped over the hedge and settled on the hedgerow some little distance in front of me. When I caught up with it it flew off and settled a little way in front of me. It did this a number of times as if escorting me on my last short stage of the journey. Now I had turned into the drive, scattering the sparrows that bathed in fine flint dust on the farm road. A familiar figure was running down the garden path smiling a welcome. My nervousness vanished as I felt a warm glow surging through my veins. It spread over my face. My eyes sparkled and with a cheerful laugh I said 'Your wanderer has returned.' I kissed the cheek of my mother. Her face which had looked so worried and worn with care when I had first joined now rejoiced in a warm and tender welcome. I stepped back to let the whole family look at me and I felt a thrilled and worthy son, for there is no greater feeling than that of being a hero in the eyes of your family, even if you are only a number on the records of His Majesty's Navy.

The fortnight had gone. I had cycled around the roads, visited the village, joined in the games and been in the village pub, where I had produced the piece of rock gleaming with silver and gold which had come from the mountain range in the Andes. One or two of the old men had gazed at it rather vaguely and remarked 'I suppose there is gold in them there mountains', and had relit their pipes, remarking on the weather and the fact that we needed a rain on the crops to make things grow, as if the Andes and South America were quite beyond their comprehension. A rain to them was much more realistic and worth more than the gold in the mountains of the Andes. Rather crestfallen I had quietly put it back in my belt. The young ones too had looked at it, still talking of their coming match on the Saturday, then handed it back without a word. I had not grasped the imagination or touched a spark of wonderment in anyone. They seemed engrossed in their own private world, where they reached the peak of their excitement and earned their bread and butter.

    I did get some consolation from one old lad who had been in the navy for twelve years, and there in the corner of the old tap room with a big frothing pint he treated me to, we went over all the old routine on a ship. Did it still exist? Had times changed? He unfolded to me his marine career as if delighted to have a good listener who understood something of the life he had endured, and from then on, whenever I was home, we would treat each other to a pint of the best and I would regale him with an up-to-date account of my life as a sailor.

I attended the Saturday night dance and again met the sweet blushing country maid who held lightly on to the arm of her future husband as together they danced the waltz or swayed in hand to the veleta. The more daring ones tapped, swayed, twisted and turned to that lovely modern dance, the Charleston. The village band boasted a grand piano, played by the schoolmaster, a cornet, played by the blacksmith, a violin, played by the carpenter and a farmer's boy who thumped on the drums and crashed the cymbals at the end of each dance. The maids with bright eyes and glowing cheeks would gather in little groups and the boys walked back to their pals, or gallantly purchased some refreshments for their last partner, perhaps hoping their generosity would further their prospective advances. The middle-aged mothers helped with the refreshments, or fluttered around selling tickets for the night's draw. The generous landlord of the local had given a bottle of whisky, a farmer's wife had given an elaborately decorated cake made with fresh farm eggs and home-made butter, and iced by the girl who worked in the bakery. Everything combined in one form or another to keep the social life of the village going.

I had been greeted and asked how I enjoyed naval life. I had walked home from the dance with one of the young girls, through the dark country lanes, only to discover I was still shy and timid, despite my bold experiences of South American night life. Somehow I had the feeling I was getting out of touch with the everyday happenings of the village. After over a year away a small gulf was beginning to open up. Maybe it was my fault. I could not explain clearly enough my adventures abroad, or describe the comradeship on board ship and naval life. The village had its comradeship, its adventures had been going on for hundreds of years. Why take much notice of one farmer's boy who had departed to sail the seas?

CHAPTER TWELVE

I walked back into barracks and handed in my leave ticket, walking boldly up the road where stood a large figure of Nelson. The undaunted courage captured in that carved statue made me feel that life was now beginning, and adventures lay before me.

I must start my gunnery course. Happily I mounted the steps to enter Anson block, commencing my second term of barrack life.

I have been taken over by the Gunnery School. Belt, gaiters and rifle are my equipment. I am drilled, taught and expected to give my whole attention for three months to gunnery. At the end of that time, if I pass, I will have a badge to sew on my left arm. A red badge of a gun muzzle with the letters 'S.G.' (Seaman Gunner). Red badges are for my number three suit or everyday working clothes and on Sundays my number one suit will be decorated with gold badges and I will draw an extra 3d. per day in my pay. 'It is not the amount of money but the honour that goes with it,' was the comment of our instructor. 'We want to take over your bodies and put the shape back into it where it was first intended to be, not where you have let it sink to round the middle – it has got into the shape of a brewer's barrel.' He walked up and down our ranks prodding us with a bayonet, saying as he pressed the point into our middle whilst we breathed in to reduce our bellies and to expand our chest, 'That's how I want you to be when you have ended your course. Your training starts from now, and everything you do must be done at the double to get you into some kind of shape in the quickest possible time.'

A series of loud, prolonged orders followed with a snap at the end. At that instant it had to be obeyed. All morning up and down the parade ground, 'Right turn. Quick march. Squad halt. By the right – dress. Stand at ease. Stand easy. Don't relax until the order "Stand easy" is given.'

Some semblance of order soon began to make itself shown in our ranks. After a few days the stiffening joints became flexible, the hardened muscles capable of throwing the rifle across the body and maintaining steadiness, so the line was kept from swaying during the minutes of silence demanded when on extra-special occasions the men of the Gunnery School were called upon to provide a guard of honour.

This then was the beginning for me, this thorough training which I was now being given. I did not know at the time for what it was intended, what I should be called upon to do for my King and Country in both peace and war. I concentrated enough to pass on from one subject to the next. Now that drilling had developed us into some semblance of the shape and form required by the Gunnery School, more attention was given to our brains, to sharpen our intelligence, to try to make us understand the working of the guns. If the whole team did their bit and each man was an expert at his job, a naval shell would travel through the air to hit a moving target over 20,000 yards away. All this then was part of my course, the drilling on parade, having some knowledge of each type of gun, its working and loading to deal with aircraft or surface attack. To be efficient, alert and ever-ready to go ashore as a landing party with machine guns and rifles. This was the gunnery side which I specialised in.

The other side was, of course, the everyday seamanship that I had been taught, and I had a year's sea time to my credit. I was becoming a member of some importance, someone the Drafting Officer could put his finger on when commissioning a ship, and requiring a Seaman with gunnery knowledge.

'This strict training is all very well,' said Ginger, the big broad-shouldered boy in our group, 'but I'm letting it get too much of a hold on me. I'm doing everything by numbers. When I went home this weekend I was even doing my courting. There we were walking up and down with me saying "Left, right, left" as we went up the street, "Change arms", and I would pass her across my body from one arm to another. She got hopping mad and finally she asked me if I was taking her for the rifle which I never left off talking about. I'm going to relax a bit and get it off my mind.'

'Good idea,' we all agreed, 'too much of one thing makes you narrow-minded and a bore to others.'

I was plugging away as best I could during the evening hours, going through the notebooks, trying to memorise all the written instructions which we had to copy down. I needed some relaxation or something to take my mind off gunnery too. No use going to the pub whilst on the course. I could only afford to go home once a month on my long weekend which was from Friday 16.30 until 08.00 Monday morning. The other three weekends were spent one on duty, and two in a large almost empty barrack room with the few who like myself lived too far away to afford the railway fare too often. We sat in the mess talking of the best way to spend our time. Should we go to the pictures? What was showing? Was it best to step ashore or return to barracks? Two North country boys talked of their girlfriends, two sisters whom they had arranged to meet on their evening off. 'Haven't they got a sister?' I asked. 'I'm doing nothing.'

'Yes, they have,' was the answer.

'Just fix me up with a date,' I said.

'Well, they did mention something about it if we had another friend. These girls have come from Canada and have only been over here a couple of years.'

'Canadian?' said I, jerking my head up, 'three sisters? I believe I have met them. Do they come from Gillingham?'

'Yes, they do.'

'I'm doing nothing the whole weekend. See what you can do for me.'

For some reason I had become more alive. I spoke earnestly and reminded them not to forget. The vision of a young girl running in and out of the crowd, a challenge that existed in her teasing eyes still lingered at the back of my mind, and now made me sit up as if it had awakened something. Then I remembered those curls that hung a few inches down her back. As she ran, turning her head, her curls swept round partly hiding her face, and from underneath the brown curls, two hazel eyes, soft and smiling, had haunted me ever since. Now more revealing than ever, as if surfacing from the depths of my memory for the first time. I rose quickly and walked away. I shook myself, walked back to the mess, got my notebook out again, looked at it, but read nothing that registered.

I tried to forget, only to be shaken late on Friday night by my two friends.

'I hope you don't mind us waking you up, but we thought you would like to know we have fixed that date up for you.'

'Have you?' I whispered.

'Yes, you have got to be by the Railway Station at 2 o'clock on Sunday afternoon.'

'Gillingham Railway Station' I repeated.

'Yes, two o'clock Sunday afternoon.'

'Pipe down there,' ordered the sentry, 'no talking this time of night.'

I lay back in my hammock, a smile of contentment on my face. Sleep came two hours after.

At last it was Sunday afternoon. I brushed my shoes twice. I straightened my silk, made sure my lanyard was just right, looked at the clock, brushed my hat, putting it at just the right tilt. It was my best collar to go with my number one suit. Now I was ready. Five minutes to get down to the main gate, then I would allow myself half an hour to walk to Gillingham Station. After what seemed an endless day, I was on my way, trying to walk calmly and unhurriedly. I handed in my leave card at the window by the main gate, walked out, turned left and took the road to Gillingham.

The green fields and the hills on the side of the river seemed fresh and beautiful in the warm spring sunshine. 'What a wonderful day,' I thought. 'Will it be one to remember or will it be a day I wish to forget?' I looked at the clock on the tower of the Officers' block. I had fifteen minutes' walk before me and twenty-five minutes in which to do it. I must waste ten minutes looking in shop windows when I reached the High Street. Strange that I seemed to be the only one walking. No one else was on the road so early in the afternoon. Most families would be having their dinner. I rounded the bend at the top of the hill. Another five minutes and I would be in the High Street. I tried to slow up but it was no use. I was beginning to feel uneasy, wanting to walk faster. Very soon I reached the High Street with the clock saying fifteen minutes to go. I turned and walked back, stopping to look in shop windows, turning every few minutes to watch those clock hands and counting the dragging minutes. Two young girls went by and giggled. I started and turned uncomfortable. Were they thinking me funny as I was looking in a butcher's shop window or were they trying to attract my attention? It made me walk up to the clock for the third time, then turn and walk back up to the end without once looking round at those clock hands. I felt I had a grip on myself now. I turned for the last time to walk that High Street. The station was just on the right at the other end. Will she be waiting? I bet there isn't anyone. Why did I take everything for the truth? My two mates may have been pulling my leg. Only another hundred yards to go and this was the first time doubts had entered my head. What shall I do if no one is there? Shall I wait? I kicked the pavement as if I didn't care then looked to see if the road was clear and crossed it. I held my breath as I looked up to the small courtyard of the station. I saw a light coat. A young girl stood there. I had a dozen yards to go. I dared not look again. That first glance was enough to set my mind at rest. Shy and timid I approached, not knowing what to say or do. My legs seemed to drag, my lips were dry. Now I had to look. A vision of loveliness stood before me and something told me I was not the only one who was shy.

'I've come to meet you,' I stammered out.

'Yes,' she answered, ever so softly.

I felt awkward and wondered what to say. All the other fellows were never at a loss and always had a ready answer. Why did I have to be so dumbfounded? Then I heard myself saying 'My mates made arrangements with your sisters. Is there anywhere you would like to go or do this afternoon?'

'We will take a walk. It's a nice afternoon.' Together we were moving away, step by step, leaving the station to cross the road leading up a long street. Together we must have walked a hundred yards. All that time I was trying to think of something to start the conversation again. No nice romantic words came into my head. The only thing I could think of was a remark about the weather, saying 'Thank goodness it's not raining.' Then kicking myself for talking about rain on a nice sparkling hot day. Wasn't it just the dullest thing to say to a girl. Maybe she was thinking I was as dull as our English climate. 'It does seem to rain on most of my afternoons off, but today has turned out lovely. Maybe it's because we two have met,' and I stole a side glance, becoming a little daring in my shyness. I saw that strand of hair hanging over her forehead and the tip of her nose. Then I looked up the street.

'You don't mind a long walk, or perhaps you would like to go to the pictures?'

'No, the afternoon seems more suitable for a walk.' With this remark I felt more at ease and quickly replied, 'I think so too. Do you ever remember meeting me before, over a year ago?'

'Yes,' she replied, 'down the Strand, when you were only a sprog dressed up in that ill fitting uniform that you wear when you first join up.'

'I've been abroad since then,' I hastily replied, 'and thank goodness now I have a tailor-made suit.'

'Oh, I don't mind the other suits so much. It's rather amusing to see all the different figures dressed in their over-sized suits, tall, short, fat and thin. You all have to make a start. Leaving home is a bold move. I wish that I had a brother, but we are all girls.'

Now as we talked and turned to look at each other I saw the puckered brow as she spoke in earnest, her hazel eyes sparkling, her sweet smile, her wisps of hair, the curls that hung round her neck. We walked along a path high upon a bank with the afternoon sun blazing in the blue sky, but we did not see it, for our eyes were gazing into each other's in the first real romance of two young lives.

We ran down to the bottom, then challenged each other in a race to the top of the most difficult part. Then I helped my young companion up the steep and slippery bank. Our eyes were shining with life and fun as panting breathlessly we reached the top and our first shyness was now partly broken. I gasped out 'You did so well it took all my energy to keep in front.' I gently let go of her hand as she smiled a soft gentle smile that already had some hold over me.

We silently sat down together, alone in the warm peacefulness of the afternoon. I watched a far-away white cloud on the distant horizon with the sunlight flowing on its edge. All the rest of the sky was a deep blue. I felt so happy, and at the same time something stirred in my inner thoughts as if I were no longer alone. I felt I had discovered something in my life I wanted to share in order to make it complete. Something was happening inside of me.

'It's getting late. It must be time for tea.' At the sound of her voice I jumped to my feet. She rose quickly, not waiting for my proffered hand, and together we retraced our way back to the town, down the so-quiet streets of that Sunday afternoon. We walked along and reached the corner of the lower road. I must have the courage to ask her if I can meet her again. We did not stop and she led me along a back alleyway.

'You must live close by,' I called out, anxiously wanting to make arrangements for another date. Several paths led off the alleyway to houses. Would she suddenly disappear down one of them leaving me stranded? Alarmed, I saw her opening a wooden gate. She stopped and looked at me as she held it open, saying 'Come in and have some tea.' She stepped off down the garden path, leaving me to follow with some distance between us as I hung back in a rather timid manner. I did not feel I was a captured treasure to be displayed in front of the family, and I felt more like turning and running. She had now reached the back door of the house and was standing on the threshold holding it open.

I was led into a room and boldly introduced as her boyfriend. Her parents accepted me by saying 'Find a chair and get in somewhere.' Looking round, I saw my two chums already seated and two girls very similar to look at as my afternoon companion. A chair was quickly forthcoming, and I asked my two Navy chums if a slender craft could berth between two old ships.

Now seated, and having had time to overcome my rather bashful entrance, I looked around to see Mother and Father and their three daughters in pretty spring dresses. I placed my tea carefully on the table, frightened of spilling it. Finding myself in so much feminine company made my hand shake, and each time I looked across the table into the eyes of my girlfriend, I caught my breath, looked down and fumbled around with my bread and butter. I was thankful my two friends were much more talkative than I, and I kept politely quiet letting them talk of their experiences, and their home town in the North country, contented to be a back number.

Some time passed whilst we smoked a cigarette and passed the time away talking between ourselves until the girls returned to the room asking us what we would like to do. After some discussion my two chums departed with the two sisters, leaving me with my girlfriend and her parents.

'Would you like to come in the other room and look at some photos,' she asked, 'before going out again?'

'Yes,' I replied, rising to be shown into the only other room, their sitting room.

'Sit down in that chair.' She pointed to a large armchair. 'I will get out the family album and show you the family out in Canada.'

Taking a large album from the bookcase she seated herself on the arm of my chair.

'Were you born in Canada?'

'Yes, right down in the south of Saskatchewan, one of the wheat-growing provinces of Canada.'

'Tell me all about this country,' I asked, 'what work did your father do? Did you live right out on the prairie?'

'Look,' and she pointed to a small photo, 'that's the house I was born in. My father built it himself, and this is the railway track running by. In this one you can see the engines, so much larger than your small engines over here. We used to wave our handkerchiefs and some of the passengers would wave back and the engine driver would blow a farewell blast on the old train whistle as it disappeared down that long track. It was the Canadian National Line from Regina to Montreal – over three days and nights of travelling. My father looked after a length of this line.'

I listened in silence for into her voice crept a warm note of affection for this homeland of hers. Her eyes held a deep far-away look of happiness, and for a minute she lived again in this home by the railway track listening and hearing the whistle of the train calling to her.

'Do you love your Canada,' I said, 'one day perhaps we will go back there.' I had added 'we' quite unconsciously, for I too had lived with her in that moment of emotion that had carried her thoughts away. She got up from the arm of my chair, looked out of the window, then replaced the album. I thought speaking of her native country had distressed her, but now composed she turned to me saying:

'The evening has passed much too quickly. I must be home again by ten o'clock so if you are inclined, we will take the path round the river to the Strand, the place where we first met.'

We left the house by the front entrance, and walked along, closely aware of each other, but too shy as yet to hold hands. We came up to the river just as the street lamps flashed on. 'If you are not in too much of a hurry,' I said, 'let's sit down on this seat and watch the old river go to bed.'

In the shaded light of a street lamp underneath a shadowy tree an empty seat had presented itself. She stopped, hesitatingly. 'I don't really think I ought to. You are as yet a stranger to me', but I caught the little note of teasing in her voice as she continued, 'perhaps I can spare a few minutes', and she sat down at the further end of the seat and I at the other with a yawning gap between us. For what seemed to be a long time I could not think of a word to say, and the seconds ticked by in silence. I saw her shadowy outline in the dim light, the fringe of hair streaming out, fanned by the river breeze. 'It's nice and quiet,' I spoke at last. As I spoke I edged nearer, so as to close the yawning gap between us.

'Tell me more of your life on the ship when at sea. What is it like?' The appeal in her voice brought out her slight Canadian accent which sounded so enchanting to me.

'The sea has an irresistible call', I answered, 'that only sailors hear, with a pull that draws one ever closer.'

Now there was no gap between us and my arm was stealing round the back of the seat, but now I hesitated. The true ring in her voice had held something which I did not want to spoil.

'You too must have sailed on the sea,' I said.

'Yes. It was great but I was sick,' and she dropped the ring of her voice as if she were blaming herself for being a bad sailor. 'For one day and night I kept to my cabin.'

'Don't blame yourself,' I replied, 'Nelson was seasick. Why even I was when we did our speed trials. I felt so ill I even groan now at the thought of it. You mustn't let such feelings as seasickness worry you. If we all did that, more than two-thirds of the Navy would be trying to buy themselves out of the service. Come on, let's talk about something else. It's the kind of night the mermaids love.'

Now my arm was round her shoulder, and with the other hand I was pointing at the moonbeams, and watching her hair being whisked by the warm breeze. She looked so lively and sweet to me that I was tempted to draw her lovely head upon my shoulder, where it would look so nice resting on my blue collar in the silver moonlight. 'You are more beautiful than any mermaid.' I spoke softly, and gently pressed her shoulder ever so slightly, for now my other hand sought hers to hold it lightly. Close together in the quiet of the peaceful evening I felt a longing stir in my breast. I felt I had found something I never wanted to leave. For some minutes we remained so, and I turned to look at her sweet face. 'This shoulder was made for your head to rest upon for all the rest of my life,' I whispered. I was so overcome by feelings that my voice faltered. At last I heard her voice say softly, 'It's nice and comfortable here, but the moon is having an effect on you.'

'It's not the moon,' I replied, 'it's you. I have a feeling I'm losing my heart and you are stealing it. Ever since that other Sunday afternoon nearly two years ago I have been troubled by a memory. Now at last I have found you again. Doesn't that mean something?'

She stirred and sat up. 'It means just this my sailor boy,' she answered, 'that it is time we removed ourselves from this seat.

Time waits for no one, not even two young souls falling in love,' and she jumped to her feet.

Suddenly all my romantic feelings seemed to descend to my feet as I slowly arose.

'Come,' she spoke joyously, grabbing my hand, 'let's hurry. It is time I was in and we have twenty minutes walk in front of us and there will be another time in the future for you to continue.'

'Do you mean that I can see you again?' my voice became more eager.

'If you want to,' she replied.

'Want to! Of course I do. I was trying to tell you back there,' and I turned round, looking back as if regretting leaving the seat. She only hurried on, however, leaving me some yards to catch her up.

'Oh, I thought you had turned back,' she said, turning to me as I reached her side again.

I could see as the pale moonlight played on her face that saucy smile again, for I was beginning to feel annoyed that my genuine and sincere feelings which I had so seriously tried to convey had meant nothing to her. But now, looking at that smile again, my anger melted and something told me that I had a priceless gem by my side and in my heart I knew I did not want to lose this treasure.

As if to put the doubts and fears out of my mind, and to make known that I had made some impression she said 'I have enjoyed a lovely evening with you.' She took my hand again, swinging it to and fro as together we silently walked the rest of the way up the street. Turning the corner, she stopped. 'This is as far as we go. I work here at a doctor's, living in. As you have already made me ten minutes late I must say goodnight.'

'When is your next time out? When can I see you again?' I asked.

'I'm off Thursday afternoons from 2 o'clock,' she answered.

'The first liberty boat is 16.30. I can see you by 5 p.m. on Thursday, Where shall we meet?'

'The same place.'

'5 p.m. by the station,' I repeated, still holding her hand. We had slowly reached the gate, and she had taken one step backwards into the pathway.

'Have I to leave like this?' I asked as I held her hand more tightly to draw her closer, looking clearly into her eyes. She turned her head slightly, putting a finger to her lips. 'I do not give my rewards lightly. Patience is a virtue.' She withdrew down the pathway of the house.

'Goodnight,' she answered.

The door closed, leaving me standing looking at that closed door as if I couldn't believe she had gone. At last I slowly turned away, walking down the street in the direction of the barracks. My head was held high. Feverish thoughts of romance filled my mind and my heart beat fast in a breast that was overflowing with a strange warm happy feeling that I had never felt before.

All next day I would jerk myself back into reality, to try and follow the words of our instructor. For minutes my mind would leap into a romantic dream and the voice of our teacher would fade away.

As our class sat in a semicircle round the instructor, who was pointing to different circuits shown on a large diagram slung over an easel, following lines of what took place when someone pressed the trigger, I subconsciously heard my name and the voice repeat it. 'Yes, you. I'm talking to you. Come out in front and just show the class what I have been telling you. Here is the pointer, just go over it.' He stepped back, looking at me, for I could not remember, nor had I heard. All I could say was 'I'm sorry. I could not follow what you were saying.'

Then from my class, a really chummy class of friends came a chorus of song: 'The love bug will get you if you don't watch out.'

Red-faced and confused I stood there, realising my closely guarded secret was only too obvious to the rest of my classmates.

'Go and sit down,' said the instructor, 'and for Lord's sake pay attention in future. Otherwise you will be on the carpet. I'm not wasting my time while you sit there woolly headed, dreaming of a bit of skirt ashore.'

Slowly each day went by, for now each day was counted. We now met twice a week, sometimes three times if our times off duty could be so arranged. Our friendship was fast developing into courtship. Even weekends when I would normally have been travelling home I booked a room at the Navy Club, so that we could spend the whole of Sunday afternoon together. We roamed the bylanes and footpaths, toured the High Streets of Chatham and Gillingham, having tea in the café or sitting together in the cinema with her head resting on my collar. There were no more bashful meetings, glancing looks or stamering words. Now we held our heads high. I could pick her out on a crowded pavement some distance away and both of us would be smiling on seeing each other.

At our second meeting I asked her if I could call her by her Christian name. I knew that her people called her 'Belle'. 'My one and only Christian name is Isobelle', she had replied.

'My Isobelle,' I had said instantly.

'And your own?' she asked.

'Just Sandy will do, but Reg if you want my true name.'

'I will call you by whichever seems most suitable at the time.'

My instructor now found me a pupil who paid much more attention to his lectures, and the far-away dreamy look was turning into a keen brightness born of inspiration.

From home a letter arrived saying my parents wished to see me as soon as possible. They had been approached by the representative of a brewery firm who had offered them the opportunity of taking over as landlord one of the village pubs. They wanted me to come home as soon as possible to look into the matter further. I explained to my young lady the necessity to go on weekend leave as they wanted to consult me before making a decision. It would mean a big change in my parents' rather settled country life.

I had tenderly kissed my sweetheart goodnight one Thursday evening under the garden wall in the small part which was high enough to shade the pavement from the street lamp. It was there that we lingered in each other's arms for the last few minutes left to us, making arrangements to meet again and sealing them with a last embrace.

'I have to go home for this,' I told her, 'and I won't see you again until Tuesday evening.'

We had counted each day we would be apart on our fingers. 'Five whole days before we meet again,' she whispered as we stood with arms round each other, 'the longest time so far we have been apart.'

'And you,' I said, 'won't go out with that soldier boy whilst I'm away.'

'Of course not,' she answered so sincerely that I planted a small kiss on her cheek, saying that I was sorry I had asked such a question.

'I will be thinking of you all the time, and Tuesday can't come fast enough.'

I had finally kissed her goodnight and she had stayed by the door giving me a little wave of her hand as I started my long walk back to Depot.

CHAPTER THIRTEEN

Friday came round and I hastened away. It thrilled me to think I was going home, but I was really in love and I knew I wanted to get family business over and get back. Quite late in the evening a rousing knock on the door of the cottage announced my homecoming.

I took my favourite seat on the couch under the window and looked round all the faces in the dim light of the oil lamp that lit the room from the centre of the sitting room table. Then we held a family discussion about a possible change which, to a poor working-class family, was the biggest event in the whole of their lives.

'Aren't we quite content as we are?' said my Mother. 'All the family are off our hands and we are now able to have one or two little extras. We have a steady job and regular income. Who knows what will happen with so much unemployment? Your father may be out of work as the pub won't keep us. We have been told that.'

'I happened to mention that to keep a pub would just suit me', said my father, 'and next day a representative of the Brewers came to see me. I only spoke in fun but the landlord must have thought I was serious. It was the other Saturday night when I went down for my weekly drink. I don't care what we do.'

'It's like this,' spoke my older sister as if she were already endowed with wisdom in her youth, 'why stop in this outlying place, two miles from the nearest village, living in a cottage which the farmer will throw you out of when you get old and can't work? Then it's the workhouse. Let's take it. It is the one chance to better ourselves.'

'Can we afford it?' I asked, as I was always the thrifty one, being a bit too careful with my money according to the rest of the family.

'I'm not going if it means we shall be in debt,' cried my Mother. 'That I will not.'

The enthusiasm of the children being balanced by the dread of ever being in debt of our ever-careful mother, my father leaving it to us to decide, we eventually came to an agreement that if we could pay our way and have £10 over she would give her consent and my sister and I were to visit the place and find out the costs. This we did the following day, and came back with the news that it was well within our means and there would be enough money over to purchase the first week's load of beer. My mother would not be in debt.

Cheerfully I returned from my weekend, having been given a happy send-off from the station by my sisters, who said they hoped on my next long weekend off they would welcome me home from behind the bar of the 'King's Head' with a frothing pint of good old English ale.

My gunnery, courting and change of home kept my head busy and my heart full, whilst I patiently waited for that letter. The gunnery course was almost completed; only another week, and that was to be spent at Sheerness Rifle Range. Sheerness was a small seaside town on the Isle of Sheppey at the mouth of the Medway. My girlfriend said 'It's just right. I will come up on Sunday for an afternoon by the sea, bringing my bathing costume, and as I can't swim, you can teach me.'

'I'll do my best and won't let you drown. I'm only a moderate swimmer,' I had told her. Our class boarded a small launch at Chatham Dockyard, taking our hammocks and the necessary baggage for the stay.

Sheerness is about twenty miles down the winding river: it made a pleasant break after over two months of barracks and I fancied myself as a rifle shot. I thoroughly enjoyed it and finished up by getting a high enough score to entitle me to wear the coveted marksman's badge of two crossed rifles on my forearm.

One incident showed how good and true our aim was. This range of rich green grass was kept close-cropped by a number of sheep. One of these luckless animals strayed from the rest during our exercise and crossed our line of fire. Maybe we mistook it for a moving target or maybe a stray shot accidentally hit it, but before the barrage of shots had died away the poor sheep had somersaulted over and now lay still, even before the firing officer's alarmed order of 'Cease Fire' rang out, and 'Unload Rifles' was hastily given.

'What happens now, Chief?' asked the Officer in Charge.

'Leave it to me, Sir', spoke up the Chief who was in charge of the range.

'I will get it removed, and there is a rating attached to Staff who has been a butcher, Sir. He can dress it.'

The dead body of the sheep was dragged clear and the butcher got to work, hanging up the animal and bleeding it in a thoroughly experienced way.

'It seems to me rather strange that no hit was registered on the target in that last rally,' remarked the Firing Officer, 'I will see this class at the end of the day.'

After several minutes' wait up stepped the Chief. 'Now,' he barked out, 'who is the villain who shot the sheep? Do you know there is a law still in force that you can get hanged for killing a sheep? Not that we shall carry that punishment out, but there is one amongst you who speaks the truth when you all deny aiming at the poor animal, because the butcher reports instant death, nine hits out of ten shots ruining the poor animal's skin, which makes it less valuable on the market. As punishment you will bury the offal and be on mutton for your dinner tomorrow. I've spoken to the Officer. He is letting you off this time. Class dismissed and don't shoot any more. Not this week', he murmured, walking away.

Five minutes later, after someone beckoned us to look out of the window, there was Chiefie with the butcher with a large paper parcel, carefully putting it in the boot of the Officer's car and finally putting a similar parcel in his own cycle basket. 'Oh well,' someone remarked, 'we won't be the only ones on mutton tomorrow.' I have a sneaking suspicion that's not the first sheep that's been shot, by the expert way that butcher got to work.

Thankfully we returned to barracks after the weekend, for we had fried mutton for breakfast, mutton dinner, followed by a mutton chop with gravy for supper so that we all declared that if we didn't get away we would be growing a woollen coat. On arrival at the barracks there was a letter awaiting me telling me to hurry and come home as my father was now the proud proprietor of the 'King's Head', drawing pints of ale for weary and thirsty villagers.

The following week we passed our final exam and I became a Seaman Gunner, proudly sewing my gold badge on the sleeve of my jumper, carefully making sure it was at the right level, no elevation or depression showing as I pulled my jumper over my head, looking in the mirror to see if it was just so, straightening myself, now half turning, for this was my first badge and I was fully conscious of its presence. I walked back to my classmates who were all doing likewise, getting ready for the weekend leave, only I was ready for night leave. No longer would my young lady be walking up the street with a young sailor showing no badges, that little badge of gold meant so much to me.

'How is that?', I asked them, 'Is it straight? Are you sure?'

After finally being assured that it was, and feeling satisfied myself, I was away out of the main gate, swinging my arm more boldly as I walked up the Gillingham High Street amongst the rest of the sailors, P.O.'s with crossed anchors and three stripes, Leading Seamen with anchor and perhaps one or two stripes, and others like myself who, until today, had nothing to show on their arm.

So with the extra confidence one feels when one gets a higher step up the ladder of success, did I that evening meet my young lady, bashfully feeling aware of her gaze as, looking at my arm, then at me, she said, 'You have got your badge.'

'Yes,' I answered, pleased that she had noticed right away, 'I am a seaman gunner with threepence a day rise, which calls for a celebration.'

I took hold of her arm and wheeled her round. 'This calls for a high tea. We will go into a café and order something nice, then go on to a show. No more swotting night after night.'

When we were sitting at a table in the café, she remarked, 'I am already getting to know your likes and dislikes.'

'My weaknesses, you mean', I said.

'Have you any?', she asked coyly, 'I've yet to find them, not only weaknesses but fears as well.'

'Do you know what this means?' I asked, turning slightly serious in my tone of voice.

'What does it mean?' she asked.

'Just this,' I replied, 'that in a few weeks' time I will be drafted to a ship and have to leave for maybe two and a half years commission.'

'That's a long time,' she answered.

'It is,' I said, 'I can't expect you to be waiting for me all that time, can I?' I looked into her eyes, feeling deeply moved at the thought of going away for so long and I was searching for an answer, not so much in words, but right into the interior of the person, perhaps in their very soul, when so much depended on what you see lying in the heart of one who can make or break your whole life. How could I expect a girl to wait all that time living in the centre of a naval port?

'Of course you can't expect me to wait so long for you. The moment you sail away, I shall be meeting my soldier boy,' she answered, holding my eyes in a steady gaze. From my thoughtful fears and serious thoughts I suddenly changed mood and gave her a loving smile, saying, 'You have everything that makes me love you. For if you had told me you would wait, I wouldn't have believed you. You just have a good look round and if you find someone better, marry him before I get back or there may be a murder, for I shall certainly challenge him to a duel.'

That night as we stood in the usual shaded part of the wall saying goodnight I whispered 'You don't really believe those stories of beautiful grass-skirted maidens? The only ones I've seen are ugly, dark-skinned natives who don't talk English and run away when we go ashore. You, compared to them, are a priceless treasure that I have locked in my heart.'

'Perhaps you may unlock your heart and drop the key overboard,' she whispered.

'When I leave', I said, 'You shall seal it with a kiss and hold the key until my return. I have a habit of mislaying keys.'

'I don't believe anything would be locked up for so long. You just be your ordinary self. No such romantic promises,' she said, 'now be off with you down the road. It is past my time to be in', and she slipped out of my arms through the gate, closing it as she threw me a departing kiss.

Back in the barracks routine, we waited for draft to a recommissioning ship. This is a kind of waiting period that suits the local married ones, but the roving spirit of the keen young sailor was restless. For myself, I felt I was between the two. I was thrilled to think that very soon I would have another ship, but I was dismayed to think I would be leaving behind the girl I loved.

At home, all seemed to be going well. I had been home for a weekend to see the whole family busy cleaning and scrubbing and serving pints of beer. I was struck with the possibilities that went with this village pub. There was a small farmyard with stables and barns and a large front courtyard making a first class pull-in for thirsty travellers, a good-sized garden with a small field at the back, and to the side a small meadow which was just wide enough to hold the village Fair, which was held once a year on the first of May.

Even in my small village I noted the changes that were taking place. More motor bikes, more cars, were seen. Tractors were rapidly driving the horse off the farm. The village cricket team no longer tied their bats and pads on to their push bikes and cycled to the neighbouring village, but hired a twenty- or thirty-seater bus, filled it with their loyal supporters and played teams much further afield. Daring fliers, both men and women, were crossing wide oceans in solo flights. Gradually the lives of the poorest people were being lifted up by the educational opportunities which made it possible now for sons of ordinary working-class men to go to University. Young men no longer relied on the big landowner for employment; modern transport took them into the town to work in factories. Political battles raged between the two extreme parties, the Conservatives and the Labour party. The once-powerful Liberal party was slipping fast down the ladder, and its leaders had only the power to sit back and make an occasional speech, while in the international field, of which now I felt I had some slight knowledge over my less experienced countrymen on account of my overseas travel, I read with growing concern the forceful and raging speeches which were now being delivered by the two dictators Hitler and Mussolini, from Berlin and Rome. The leaders of Britain remained timid and uncertain, most of their time being taken up with home affairs.

Meanwhile the trade unions, many thousands strong, were giving power to men who thought they could remedy overnight the poverty and wrong which they blamed on the others who had governed before them. I read and listened and had my own thoughts and opinions. I was pleased that I lived in this age which was giving the common man a chance to be educated, to keep pace with the ever-increasing progress that scientists made possible by their discoveries.

When I looked back and realised the many things that had befallen me in the few years that I had forsaken the land for the sea. I realised why in our school days we were taught to sing 'Rule, Britannia, Britannia rule the Waves.' I was beginning to have some understanding of what was required by a small country to keep a powerful fleet afloat, to train its men to use modern scientific equipment, and also to keep them tough and resilient.

I was only a young A.B. in this big force; at my age I was

anxious to get back to Chatham from my weekend at home to see again the girl who now occupied most of my thoughts, who seemed to be taking first place in my life. The only thing I was anxious about was whether I would be on duty on the evening my girlfriend was off. Each afternoon or evening that it was possible to see each other we would meet, spending a few hours in each other's company. They seemed to pass so quickly and the days in between seemed to pass so slowly. Love cast such a spell over our young lives. Together we waited to see what ship I would get and to which part of the world I would go, and whilst we waited we had these blissful, joyous meetings.

At last I saw that my name had been included amongst those due to join H.M.S. *Ardent*. We were to have fourteen days' drafting leave. What then, I was thinking, would happen to our courtship? Was it strong enough to last, to stretch those long years away without breaking? Would I find her letters getting shorter with longer intervals between, until finally they stopped altogether? Thus preoccupied with misgivings and doubts I went that evening to meet my sweetheart. I saw her casually looking in shops in the High Street, this being our regular meeting place.

'What are the arrangements for this evening?', I asked, 'anything special?'

'Nothing at all,' she answered, 'I will leave it all to you.'

'Come then, we will take the path that we took on our first afternoon together, for I have something to tell you when we get out of this crowded street, that will affect our lives.'

'I know what you are going to tell me.' She spoke first.

'What is it then?' I asked.

'You have a ship.'

'Yes, that's right, and it is going abroad.'

'Yes, that is so.'

'How did you know?'

'Woman's intuition, besides I can read you like a book. Your thoughts are written all over your face. The serious, thoughtful expression, no ready smile to greet me. Anyone would think the end of the world had come. Already your mind is thinking of how long after you are gone my letters will cease to come.'

I did not speak; she had been so right, and as if to sting me into answering she carried on, 'What kind of young man are you? Where is your faith in me? When your first and only thoughts are that I will be off with someone else the moment you turn your back. If that is the only impression that I have planted in that mind of yours it is for the best if this very evening we say "goodbye".'

Stunned and bewildered I gasped in astonishment. Indignation flared up in me. I stopped still, angry words ready to utter as I opened then closed my mouth. Owing to my sudden stop a burly figure had lurched straight into my back, sending me forward in such a sprawling way that I had difficulty in not falling. He swore an oath, adding 'Why don't you give a blast on your foghorn when you suddenly stop?'

I looked at him, rage darkening my face, my eyes glistening with fire, showing the fiery temper that was springing to the surface. But now, a few seconds of bewilderment made me hesitate, time for reason and commonsense to enter my head.

Now my girlfriend spoke, reproachfully, 'What a way to behave, flying off the handle when it was all your own fault. Where is your self control?'

She spoke to me with firmness, an edge of annoyance creeping into her voice. 'You are showing me a fierce temper which boils up quickly. I thought you were a steady, immovable type like the rock of Gibraltar.' She turned her head and began to walk away.

'Give me a chance to explain,' I called out. 'What were you saying to me to make me stop so suddenly? You made me angry.'

'Did I?' she said, taking hold of my arm, 'What with your darkened features and now your hot flushed face, you must have a fiery spirit. Come, let us go along the path of the high bank to let the air cool that flushed face. Let's enjoy the present. You must not worry over the future, it might never happen.'

Silently we walked on, not noticing the lovely scenery that stretched for miles. Then I broke the silence as if still simmering underneath, 'Didn't you say something about saying "Goodbye" this very evening?'

'Did I?' she replied, 'let's sit down. You tell me all about the ship you are going to join. If I said anything to hurt you it was because you provoked me, thinking I would be running after someone else the very minute your back was turned. What sort of girl do you take me for?'

Slowly, reluctantly I sat down. These were the first few words that had flared up between us and a little bird was already whispering in my ear that it was all my own fault. I turned towards her as she continued talking to me.

'You won't be going away yet. You will have some leave to come and I will have a holiday shortly.'

'Yes,' I replied, 'we always get fourteen days commissioning leave,' and I scooped up a handful of earth, holding my head down as I spoke.

'Well, that will be our first break. You will go home to your village.'

'I don't know that I want to.' I looked up, seeing a soft, gentle smile playing around her teasing eyes which instantly melted my simmering anger. My hand stopped scooping the earth and grass and the next moment she was in my arms, saying 'Let's see that angry glint come into your eyes again,' as she smiled up at me.

'I have a good mind to leave you right here. You have only been playing with me to see how angry you can make me,' I cried.

'You have a better mind not to,' she laughingly replied.

I stopped her saying any more by smothering her with sweet kisses. She struggled, pretending to free herself. I held her firmly, saying, 'You women play with men's hearts, leading them on, then melting them at your will.'

I spoke warning words in between our rapturous embraces. 'You have the very devil in you this afternoon, and I won't stand for it.' She sat looking at me, stroking my neck with her hand so that I was loth to scold her any more. I spoke with feeling. 'You are the sweetest, dearest girl in all the world.' I drew her close and kissed her intensely.

She suddenly pushed me back, springing from my arms, saying, 'Your lovemaking is like your temper, fierce and uncontrollable. Let's carry on walking to cool you off again.'

We walked along. Our first quarrel was patched up as if it had thrown us closer together. Her arm was around my waist and mine was around her. We held each other's hands, her head partly resting on my shoulder as I told her something about the ship I would be joining.

'I'm glad you will be on a small ship, doing lots of sea time. It will keep you from temptation.'

'Where in the world would I find anyone sweeter than you?' I squeezed her waist as we stopped to kiss. 'There won't be any other, only you,' I whispered.

'I don't want there to be,' she whispered back, and now we lingered underneath a leafy tree, resting on its trunk. As we clung to each other as young lovers do the daylight faded, darkness gathered and the stars were twinkling. The night wind stirred through the leafy trees as I whispered 'My darling, we must go back. It was light when we stopped, I'm keeping you out far too long. We are sweethearts now and we will wait for each other no matter how long we are apart.'

'Yes,' she answered, 'Always.' And we strolled back with our arms round each other along the path that we could only just see, back down the lighted streets to the garden wall.

'Will you come home with me on your holiday?' I blurted out.

'I don't know if I should be allowed to. You live at a pub. Have you some intention of getting me drunk?' she asked. 'What's going on in that head of yours?'

'No wicked thoughts that would harm a hair on your head,' I replied, 'I'm so proud of you I want to show you to the whole world.'

'Is that your whole world, that little village in the corner of some shire?'

'It's Cambridgeshire,' I answered, 'if it is a little village it is my home and I would so love you to come. What a lovely fortnight I would have to remember you by. Do come. You would make me the happiest man alive. Think it over, for now we must say "Goodnight".'

CHAPTER FOURTEEN

I arrive at the hotel and wander along the corridors looking for the number of the room. I unlock the door, step in and settle down on the bed to ponder over my movements until 10 o'clock next morning when I am to meet her by the station, the very station where our romance began. I go out and return with a paper and read it right through, later walking the streets, wandering aimlessly through shops, taking the long way round, spending much time over my midday meal, trying to kill time.

I am awoken in the morning by the noise of cars and buses, voices of people on their way to work. I lay and just listen, knowing it is early and I have no need to hurry. At length I wash, dress, have breakfast and am out into the street. It is still only nine o'clock. There is another hour to walk to Gillingham Station. I set off leisurely along the street. At the top of a large, steep chalk hill is a monument to all the Sailors of Chatham Division who gve their lives in the Great War. I read the inscription and some of the names of officers and men who had perished in action. Although I was alone I removed my hat in respect, for I felt these men whose names I now read were the heroes whose daring deeds had inspired me to join. I wondered what it felt like to have guns blazing at one and the sea as your graveyard. That couldn't happen – another war. I was in much too happy a mood to let such serious thoughts trouble me on this fine morning. Replacing my hat, I walked on over the grassy slope, stepping and looking round once I had reached the top.

I was first to arrive at the station. I saw the bus coming up the hill and as it went by I saw her standing ready to jump off. I crossed the road to greet her, and I gasped 'I thought this moment would never come, but now at last we are together for a whole fortnight.'

'You don't want to turn back at the last minute?', I ask her.

'You haven't given me much chance to turn back, the way you grabbed my hand and rushed me through the station,' she replied.

I smiled, saying, 'Sorry to rush you, you look so nice in your travelling costume. I'm not letting go of your hand until we are seated in the carriage. Let's look out for an empty carriage all to ourselves.'

She took her seat beside me and I reached out and took her hand. 'I love a train ride,' she said, 'it brings back memories of my childhood days in Canada. The day we left for that great ride for three days and nights when the whole family left, never to return. Now I'm getting off again with you. I wonder if each long train journey has something to do with my fate?'

At this very instant sudden darkness enveloped us for the train had rushed into a tunnel, catching her unawares. She gripped me tightly, partly frightened by the shock.

In London we took a bus to Liverpool Street Station.

We had thirty minutes to wait. Just right for a break and something to eat and drink.

The train moved off, rushing through the small stations. At length I pointed out of the window saying 'I have cycled along that road through that village. A few more minutes and we shall be at our station. We are almost home.'

As we walked along the platform together after sorting ourselves from the scrimmage, I said: 'This is Cambridge, the great seat of learning, where students walk around in mortar boards and gowns and speak with an accent that distinguishes the cultured gentlemen's sons. You will se them riding their cycles five or six abreast, their books clutched in one hand, chattering away.'

At the bus station we settled down together and the bus filled up. A bell sounded and we slowly began moving through the town, out into the country.

The old bus shook and bumped its way along the stony country road to enter the village, and from behind the thick green hedgerow amongst the fruit bushes and spreading apple trees, one caught glimpses of old whitewashed cottages with thatched roofs. Flowers grew in narrow borders, decorating either side of winding paths of cobblestones and cinders that led to front doors and trailed off round the cottages. Large elm trees with tall branches reaching high in the air and clustered with green leaves, glittered and danced, stirred by a slight breeze on this fine afternoon.

The bus was slowing up. 'This is it, our village pub.'

She followed me down, stepping off the platform, waiting while I gathered up the cases.

'This way.' I turned to her, smiling with confidence, to put her mind at ease. We walked across the courtyard. I was bringing my sweetheart home for the first time. There were introductions and the embarrassment of introducing my girlfriend for the first time. There was my mother, the centre of the home; Father with his worn hands. Shyly, but with a deeply buried feeling of pride, I introduced my parents to the girl who had now taken first place in my heart. Soon all three became acquainted and we enjoyed our tea.

'I will show you round the village tomorrow,' I whispered as I kissed her 'Goodnight' on that first evening. 'You aren't frightened now, and won't write home for your mother to come and fetch you or run back home yourself?'

'No,' she answered, 'unless . . .'

'Unless what?' I asked earnestly.

'You may have a village maid. Some sweet girl whom you promised yourself to when you sat out in the meadows together amongst the buttercups and daisies, and vowed to marry when you grew up.'

'Have no fear,' I assured her, 'You are the only one I have ever lost my heart to. Would I have brought you home if there was another? I was so shy and backward in my young days.'

'That is hard to believe,' she said teasingly, blowing me a departing kiss from the closing door.

CHAPTER FIFTEEN

It was daylight when I awoke in the morning and the sparrows were chirping quite loudly from the guttering by my bedroom window. I dressed and hurried downstairs.

'She is not up yet?' I asked my mother.

'Please take her a cup of tea and tell her the sun is high in the sky. I will go and fetch a fresh egg.'

'Breakfast's ready,' I called out, 'it will soon get cold.'

'Coming,' she called back, as if she knew of my dilemma. At her answer I tapped lightly on the door.

'Come in, did you say?' I pushed the door a little ajar.

'I don't think I did,' she replied.

'I must have been mistaken,' I answered, but entered all the same and closed the door softly behind me.

'Don't do that,' I cried, putting out a restraining hand.

'Why not?' She hesitated a moment, holding the lipstick she was about to use.

'Because I want to kiss those pure lips before you put that paint on. I don't want an artificial painted flower,' I said.

She smiled back at me. 'All this so early in the morning. Whatever are you going to be like in the evening? Your mother will be opening the door giving us a ticking off.'

'I don't care if she does,' I said.

'Well, I do,' she replied, 'It's breakfast time.'

At that moment Mother called, saying 'Come on both of you, it's all ready and waiting. Don't let it get cold.'

'In you go', said my girl, giving me a push towards the door, which I opened and we both took our places at the breakfast table.

'What is the programme for today?' I asked her after the bacon and eggs had disappeared.

'You promised last night to show me around.'

'I will, with pleasure.'

'A few years ago, seeing the older boys of the village walking out with their girlfriends, or being together at a dance on Saturday nights always made me envious. Some day I will go out into the world, I vowed myself, and fall head over heels in love with the most wonderful girl so that I could feel the same. Today I have you making it all real.'

'There is the old walnut tree by the schoolmaster's house. I've filled my pocket with walnuts many a time when the high wind has blown them off.' We turned the corner by the pond, almost covered in by overgrown branches. 'That is where we used to make a slide in the winter when there was frost. We would form up in a line, taking turns to make our run, with our hobnail boots ripping over the ice.'

'Is this the playground where you chased the girls?', she asked me, getting a word in at last.

'I suppose I did, pulling their pigtails as we ran around those trees. They looked very nice with their long plaits tied up with wide coloured ribbons. I can't remember much about girls in those days. I didn't have much time for them. I was enslaved by no one, free as the birds.'

'My poor, dear boy', my sweetheart was saying, 'you are making me feel so pathetic towards you. Do you wish for your freedom? Do you want me to release you from the enslavement of our romance?'

'Oh no,' I answered, 'One cannot go back to live in the past.' But we did walk through the churchyard with all its memories. Then we walked up the thick avenue of trees into the church porch and stopped to read the items of news posted there. We entered the church quietly and dropped our voices to a whisper. We walked the length of the church hand in hand, standing before the stained glass window, gazing up in thoughtful silence at the picture of Jesus nailed to the cross. Not a whisper was uttered and a deep impression filled my mind. My loved one was by my side; I hoped I would be worthy of her. I pressed her hand as if assuring her of my true feelings as we turned to walk back.

'Did you say a prayer?' she asked.

'Yes,' I confessed. She gently pressed my hand as if she too had asked for something in prayer.

'Are you very religious, bringing me in here? I am glad you did,' she added quickly. 'Did you have to attend church every Sunday?'

'No,' I replied, 'I was left to please myself, even as a boy.'

We had now left the church and walked round its wall to the

corner where the chapel stood, and compared the two, the grand old church and the neat new chapel. 'The chapel is a breakaway from the church, where religion springs from the priest.' In the chapel the layman stands up in the pulpit, preaching to his fellow worshippers who sing their hymns with religious fervour, but to me too much praying is like too much beer. It makes one drunk.'

'Let's wander down through the lane where the brambles are running wild. Look, see the clusters of blackberries hanging from the briars? A pity we didn't bring something to gather them in. You will have to be careful not to ladder your stockings. This is rural life. We have some five-barred gates to climb, a ploughed field, besides a meadow full of bullocks to pass through, then on to our Common with its wide stream which we village boys boastfully refer to as a river. Are you game?', I asked.

'Yes,' she replied, 'If you lift me over the gates, protect me from the bullocks and buy me a new pair of shoes and stockings if I ruin these.'

I promised faithfully to protect her from wild cattle and thorny bushes and repair any damage before proceeding down the cart track and cow trail, with its overgrown hedges and trees on each side, stepping over the deep ruts, jumping the small bushes and tufts of grass. 'It's delightful,' I cried, 'We are now in the middle of the wildest part of the parish. All alone, just we two,' as I helped her over the first gate.

'Now don't you get any wrong notions in that head of yours,' she cried, 'I believe that is the only reason you brought me down here', as I still retained my hold.

'Not me, sweetheart. I wanted you to see the beauty of our trees and bushes and the wild flowers growing all around us. That is the reason.'

'The only one?' she asked, sitting on the gate looking down at me with that teasing look.

'Of course it is,' I stammered.

'Then leave go of me', she said, 'and don't get me lost, my honest boyfriend. Get going.'

I said not a word, as I tramped along over the ploughed soil to the wide ditch with a wire fence running along its bank. I held the fence apart for my love to get through and then gave her a hand to help her jump the wide ditch.

'Wait, my silent hero,' she cried, 'will you undo the lace of my shoes so that I can shake out the loose soil?'

'Yes, my fair lady,' I answered, 'I will kneel at your feet for you to rest your dainty foot upon my knee.'

'Just replace my shoe, my foot doesn't need tickling,' she said as I carried out my task, giving her toes a slight brush with my fingers, making extra sure no grit was left there, 'and my stockings aren't laddered either,' as I said it looked as though one was starting to run up her leg.

'Is there no reward for the help your knight has so dutifully performed, even kneeling at your feet? Just one little kiss.' I appealed with a look of despondency. Leaning over, she lightly kissed me on the cheek.

'Dearest, you have now made my day,' I cried, rising up and placing my arm round her waist. 'Let us walk together between the bushes. What does it matter if we do get lost? The whole day is ours.' With arms round each other we wandered over grassy meadows to the bank of the stream and found a large bush where the grass grew lush and green. All around was quiet and still. We sat down to rest and I offered my arm for my sweetheart to rest her lovely head of hair. I whispered sweet words of love and tenderness brushing my lips gently on her sweet and lovely face, running my fingers through her curls and tresses, and felt her little heart beating rapidly as we clung together, wrapped in each other's arms. At length my sweetheart stirred, saying, 'Don't you think we ought to go?'

I smothered her with kisses, assuring her it was quite early, and there was time for her to relax in my arms. I wanted to remain that way for ever as we snuggled up close, lying cheek to cheek until my arms were numb and I had to ask to move them. She sat up, looking round.

'No need to go yet,' I whispered.

'Why not? We have been out all day. It must be teatime. What explanation are you going to give when your mother asks what you have been doing?'

'Just taking care of you. I'll tell her we got lost in the meadows and there was no one to ask the way.'

'Get up and help me to my feet, like a true gentleman,' she said, giving me a slight push to stir me to action. I reluctantly got to my feet and offered my hand to help her rise. We spent a few minutes picking the grass from our hair and shook the creases from our ruffled clothes, making sure we were quite respectable to walk home through the village. The bullocks stopped munching the grass and turned their heads to stare at us as we made our departure, climbing over the chained gate into the narrow lane on to the road home.

The days went by quickly. We worked in the garden when we felt like it, took the bus to town and went to the weekend dance on the Saturday night. My sweetheart was very much sought after. As this was her first dance, she was much too shy and timid to accept when she was asked, and only with me did she venture on the floor to try the quickstep and the waltz which we had practised at home, my sister playing the piano, teaching her the one, two, three round the parlour table. Then one evening all the young ones came in, bringing the accordion to accompany the piano. The old village pub became crowded, every room, passage and doorway. The whole family were busy, trying to supply the drinks to keep the guests happy.

'Are you enjoying it?', I asked, as my sweetheart rushed up with a tray of empty glasses. I stopped to take a closer look. 'You seem a little flushed. Have you been drinking?'

'A young man asked me to have a sherry,' she replied, 'and you told me to try to keep everyone happy. I didn't think it right to refuse.'

'Just sip a little, not too much, because when all around are losing their heads, a good waitress and landlord must keep theirs', I told her.

'He was so insistent, besides, he's good looking with nice curly hair, and very polite and helpful. Each time I go into the room he offers me a drink.'

'Beware,' I said, as someone shouted for service, and she turned away laughing, leaving me besieged with customers. Some ten minutes went by before I could get free for I hadn't seen her. Perhaps I was getting a little bit uneasy as I edged my way down the passage. I could not catch sight of her.

'Have you seen my girl?' I asked my sister, trying to be assured.

'Oh, the last time I saw her she went outside with a fair, curly-haired young man.' She flashed me a smiling glance full of meaning.

'Yes,' added someone, 'That chap with a car.'

'Oh did she,' I answered, 'Good luck to her.' I took a drink from my glass, pretending to be unconcerned, for they all gave a little titter, and one who knew me well said 'I think you've lost her already. If she is not used to drinking, a little soon takes effect. I should go and have a look outside,' and they all laughed at my expense, enjoying the predicament their merry teasing put me in. I gave a wry smile, shrugging my shoulders and trying not to let them see their bantering bothered me. The feeling that I was being disregarded caused me some alarm, and when I could I took a look through the window, and when I thought no one was watching me I stole outside, searching around without finding her, only to run into my father who added fuel to the fire by saying, 'Have you lost her? I saw her getting into a car.'

After that I quietly carried on helping, now silently subdued as closing time drew near and no sweetheart was around. Everybody was full of life, singing and laughing in a joyous way. I felt dispirited and wanted to be by myself and I slipped into the kitchen away from the clamour and the noise. My mother was laying the table as I entered, remarking 'We have had a busy night. You will all need supper. It will be all ready by the time we turn out if they get back as promised.'

'What are we having?' I enquired.

'Your young sister has gone to fetch fish and chips with her boyfriend in their car, taking your girlfriend with them.'

Immediately I turned to enter the party again, all fears washed away. At the supper table they referred to the incident and exaggerated my behaviour, telling how I paced up and down, searching everywhere, drinking pints, calling her name. I just smiled as my sweetheart rewarded me with a hug and kiss, saying 'Did I give you a big fright?'

It was the last night of our holiday and I think we both felt sad about it, as if unable to settle, for in the background lay the long parting. Was it a wise thing to do to spend our holiday together as we became more attached? Was it fair to two young lovers to spend two to three years of their youth just waiting, when all around others of the same age enjoyed all that life had to offer? I asked myself what she could see in me that would make her wait or even think I was a good prospect. Maybe it was the uniform, but there were hundreds of others in the same uniform. I puzzled over it without finding an answer.

I knew that to me she was my ideal, my true love. Love had me enmeshed in its tantalising web and I did not want to break the threads.

Alas, everything has to come to an end, and the fortnight had gone. We stood together underneath the shaded part of the wall by the house where she had to return to take up her duties. 'You have no regrets? Will you always love me?' I asked her in a voice that trembled with emotion. 'I don't want to leave you at all. Why did I join the Navy?'

'You joined it to find me,' she whispered back, 'and when you have finished your commission I shall be waiting.'

'It seems a long time, my darling sweetheart,' I told her.

'Yes, I know,' she replied. 'Remember there are thousands worse off than we are, married men with little children. They have to do their duty. How hard the parting must be for them, so cheer up my dear boy and face up to the life you chose yourself.'

'Yes, I will,' I told her, 'and when I get homesick and lonesome and am feeling downhearted and dispirited, I will think of you waiting there at home, remembering our holiday together. I will go down to the Mess and open up my ditty box. Your photo will be smiling at me as I sit down and write you a loving letter.'

'You do just that,' she said, kissing me goodnight.

CHAPTER SIXTEEN

Destroyers are little ships with slender lines. I was one of the ship's company of H.M.S. *Ardent*, which we joined late in 1935. Following in the wake of two other destroyers of the 'A' class, we would be joined by others from Devonport and Portsmouth making up the flotilla doing exercises and settling down into a happy ship's company, as our Captain put it when he addressed us. I felt so lonely in this little ship, my messmates were as yet new acquaintances. It was so different from the *York*, which was much bigger and higher. On this ship the water swept by almost level with the deck. Soon we left the smooth waters of the river, out into the grey white-flecked waves of the sea. The little ship dipped her bows, rising and falling as if in anger at being tossed about so early on her first voyage of her new commission. I stood on the fo'c's'le deck to catch the last glimpse of the mainland. Looking around I saw a gloomy grey atmosphere enveloping us as we sailed further away. The last outline of the shore appeared as a dark speck of mud. What a farewell to my native land!

My spell of duty ended and I made for my hammock. I quickly curled up, thankful to get some relief from the empty low feeling. The moving hammock swung to and fro, or shook to the ship's pitching and rolling. I lay, trying to doze off, in a state of abandonment. The mess tins rattled in their slide from side to side.

'Out you get. Show a leg, rise and shine!' The monotonous voice droned into my ears to a violent lifting and shaking of my hammock. Realisation came slowly back into my muddled mind, and I stirred wearily, with life at its lowest ebb. I turned slightly, pulling the blanket round me and over my head to shut out that unwelcome command, knowing I had still a few minutes to curl up and forget my whole existence. Next time the violent shaking was short and abrupt. The voice had the irritated snarl of the 'last time or else' about it, as the 'Rise and shine' ended with the remark 'You seasick barrack-room stanchions.'

I must face the inevitable. Catching hold of the bar above, lifting myself clear at the swaying hammock to feel the moving deck beneath my feet. Somehow I managed to pull on my trousers (which had been folded up and used for a pillow in my hammock, a useful as well as safe and handy place to put them). Steadying myself by the fixed mess table, I got partly dressed in the stifling atmosphere of the mess deck, the thick, hot air of sweating bodies. I had to move, stepping over fallen stools, ducking under hammocks, making my way quickly to the open deck, there to gulp in the chilled freshness of sea air. The light grey day, the grey sea and ship were all swaying and moving as I looked aft and out. Even the sky seemed to roll with the ship. I clutched hold of a support as I stood surveying this gloomy scene in my unwell state. I stood, not saying a word to others who had gathered around, apparently feeling as I did. A small group spoke a few words, whilst one or two passed by or ventured out as I was doing to get some fresh cool air in their lungs. The shrill whistle of the bosun's pipe sounded down from the bridge, followed by the voice of the bosun's mate, calling out the order that all hands would muster in fifteen minutes' time. I ventured nearer the side, holding my head out to get the full blast of fresh spray as the sea swept by.

Now I felt I had just enough life in me to get back on the mess deck, lash and stow my hammock and prepare to muster. As I struggled for'ard to do this, someone more alive than me shouted 'give me the boots'. I managed with a struggle to keep upright to do all that was necessary and was ready to muster when the order was piped round. Oilskins and seaboots seemed to be the dress of the day. We hung to the ship's stays by the shelter of the foremost funnel. There was no order to attention. It was 'Hang on while I report to the bridge' as the buffer remarked to us as he dashed up the ladder (fearing I think we might not be there if he left it too long). 'Clean ship' was his report on return. 'Detail extra hands for the flats and mess deck.'

There wasn't much else we could do, but even that was enough. The mopping-up operation restored some sort of respectability to that disorderly interior that had been upset by the buffeting we had received, and were still taking, but which was now more subdued. More life came into the ship's company. Messes became spick and span again. Dinner was prepared for those with restored appetites.

Gradually the ship became steadier as the sea became less troubled and the wind dropped its whining. Colour returned to those ashen faces. Men began moving about calling cheerfully to each other as they carried out their detailed tasks. The chef was now more cheerfully in control of his pots and pans, lifting the lids to dip in his ladle to stir its contents, remarking to the bystanders that this had only been a heavy sea running, nothing to what he had been through. The long-awaited call for cooks of messes to draw their rum ration came. A few more minutes then the general assembly of Jolly Tars began. Outstretched hands reached out, lifting cups to lips, throwing back their heads as each one would say 'The King, God bless his cotton socks', or 'To My old Granny, God Bless her Soul', as the cup was held high to catch the last drips. The cup was replaced and we settled down a minute for the rum to warm us and began talking as dinner was set out. Hungry men filled themselves, looked up and patted their stomachs on completion, then looked for a pillow for their heads. They selected a place to go to sleep, well pleased with themselves. The cooks of the mess took to the galley the dinner of anyone on watch to be kept hot. Their twenty-four hours on duty as mess cooks now ended, two more took over the responsibility of maintaining cleanliness, and keeping the hungry members satisfied. This was my introduction to destroyer life, where sailors became hardened seamen and gained their sea legs.

The remainder of the flotilla had joined up, and ahead lay a thousand sea miles to Gibraltar. We had to become accustomed once again to the heaving decks, the forty-five-degree roll, which had been known to increase to ninety degrees in the heavy sea of a real storm, with the yardarm of the mast dipping in the sea – or so we were informed by the long-serving members of our seamen's branch who volunteered for destroyers only. Any bigger ship would be to them like a spell of shore duty. What a difference from the *York*, which cruised from port to port all on its own. Here I was in a destroyer, in company with eight others in line ahead, greyhounds of the seas. Any second the leading destroyer carrying the Captain might hoist a number of flags from the yardarm, this being a signal which was repeated by each one, the flags of coloured bunting flying out in various shapes and bright colours. Then as the signal was run down, the whole flotilla might turn together in perfect formation, the most wonderful sight to a sailor's eye.

At last, Gibraltar appeared on the distant horizon. The flotilla of destroyers seemed to speed more easily through the sea and the waves of the Atlantic changed to a clear blue as I watched from the deck of the *Ardent*. I looked and felt on my second approach to Gibraltar as if I was a more settled and experienced member of His Majesty's Navy. I had lost the eager excitement that I felt on my first visit and meeting with my first dusky maiden. I now had responsibility resting on my shoulders. I was not the tearaway young sailor of my South American voyage, but a more serious, thoughtful type, who rationed his runs ashore, steadily saving from his small amount of pay towards the future happy home that he had in mind.

As soon as the ship was secured, all hands set about its personal appearance, scrubbing and scraping, painting and polishing. Similarly every Saturday morning was devoted to bringing to Navy perfection each flat and messdeck with each member responsible for his particular job standing to attention to answer for its efficient state, as the Captain with all his junior officers in attendance inspected and probed, looking overhead and underneath for a telltale sign of dirt that pointed to negligence.

'Has the Captain finished his rounds yet?', I was asked by Ginger, anxiously poking his head round the starboard screen.

'No, not yet, and keep that ugly mug of yours from view or we might get a rescrub.'

'I left my collar out. I want it for going ashore' was his answer, ignoring my remark about his ugly mug.

'You've had it,' I told him, 'you may depend it is in the scram bag. Can't you remember the warning of what would happen to all gear left about?'

'I wasn't there,' he replied, 'I've been ashore getting stores, detailed off on a special party. How can one think of everything?'

'The only thing you are thinking about is that run ashore and finding a Spanish señorita,' I reminded him.

'Why not?,' he said, 'ten days at sea with just a hairy matelot to look at and talk to, if I can only feast my eyes on their adorable curves again, what a pleasure.' His eyes brightened at the prospect. 'Why don't you come? Don't you want something to admire and wash the taste of salt water out of your system?'

'Malta will be my first run ashore,' I replied firmly.

'Oh, I suppose you have tied yourself to some girl back home. You'll soon forget her. Maybe the first few months while her memory is fresh in your mind, but as time increases so the memory will fade. Safety in numbers I say, I have two or three back home and I hope to have a few out here. There goes the skipper, and now for a run to shore. Take my advice and come,' he shouted back as he passed me on his rush to the messdeck as the Captain and his troop of followers left the for'ard part of the ship to finish their inspection aft. From hatchways, corners and down ladders came members of the ship's company, back into their respective messes to enquire of the leading hand of any personal remarks the skipper had made. A small word of praise coming from the top one goes a long way, although brushed aside by a sarcastic remark at the time – Jack's way of hiding his bashfulness.

I kept my word and stayed on board, and settled down to write a love letter, as my messmates threw out their disparaging remarks of 'Forget her. She is out with that Marine. You can't trust one of them. How long do you think she will write to you?' Some remarks could hurt if you were of a sensitive nature as I was, being young and in love, but I coiled up inside and remained silent. This only encouraged the backhand smiles and knowing winks.

In response to the cynical statement of an A.B., one married seaman said 'It's buying your love if you are going to send a present from every port or money each week. A present at Christmas and birthdays, yes, but true love is instantaneous. There is something that tells you the right one has come along. What it is, or how it happens, remains a lifelong mystery. Money and presents don't come into it.'

This version of the whole matter seemed to find instant approval as a few of the younger ones, beginning to get interested, wished to hear further wisdom from the married Able Seaman. 'Tell us, Stripes, do you really believe this sudden love?' asked a young curly-haired boy. 'Supposing I'm walking down the street. A girl passes. We both look at each other. Will something tell me that's my lifelong partner?'

Stripey now found himself the centre for information. He leisurely rolled a tickler (cigarette), taking his time in answering. 'Yes,' he replied. 'that is how it happens.'

Then he told us a warning tale. 'There was a young A.B. on my last trip out here. A steady type who had a girl at home waiting his return. He used to show us her photo, relating how they planned to get married and were saving up between them. They had already got a hundred pounds saved up half way through the commission. Then one day we put into Haifa. This fellow went for a run ashore with some of the others, just a usual run, for a drink or two. We passed an Arab family with their tent pitched under a palm tree. The old man was sitting cross-legged, eating dates, with his sons and daughters sedately standing by. The daughters were veiled and wrapped up. We stared, just out of curiosity I suppose, and wished them "Good Afternoon" which I don't suppose they understood. George Kimmin, the young A.B. walking nearest to them suddenly stopped, as if rooted to the spot, right opposite a young, slender, loosely veiled Arab girl. What passed between them no one ever understood, only those two, for it must have been the language of the eyes as that was all one could see of her. Poor George seemed transfixed and one of us went back, almost having to lead him away. We didn't want any trouble.'

Stripey paused, putting away his tin of tobcco. 'You never knew those Arabs,' he continued. 'We dragged George Kimmin away and told him of the danger he would bring on himself, but he was well aware of this, having been around these parts before. He came with us. He lapsed into silence and before long we missed him as he had slipped away, but this was not unusual.'

'We didn't think any more about the incident, until returning to the ship we passed the same place on the road. We looked about for Kimmin but reckoned he had gone back to the ship on his own and didn't bother all that much, but there he was, standing by a large stone, looking with a fixed stare in one direction and as we looked that way there stood the Arab girl, her slender robed figure so silent and still in the semi-light of the hot, starry evening. We called him a fool and almost had to use force to get him to return with us to the ship.

'Next day Kimmin began to act strangely. No one saw him write any more letters and he ate less food, buying lots of dates and eating them instead. He also went ashore at every opportunity on his own. He returned one night with a long roll of white cloth like the Arabs wear and the lads on the mess deck began to whisper that George Kimmin was turning native and was going to take the faith, for he had been seen again at this same spot where the wandering Arabs' tent was pitched. Rumours were flying round of George Kimmin's strange behaviour, some even said they had seen him riding on a camel, and this may have been true, becaue one evening he came out with his roll of white cloth and began to dress like an Arab with all his messmates looking on, walking up the ladder on to the fo'c's'le. His messmates stared, then followed discreetly some distance behind.

'George stood still, not taking any notice of the many faces he knew must be watching him. Then, when the Holy Priest from his high tower called the faithful to prayer, George knelt down, facing East, going through all the motions of a devoted follower of Allah. After this, he returned to the mess, put away his robes, and sat cross-legged on the mess locker, eating his dates. Now the mess was buzzing with excitement. Bets were laid that George was going to desert the ship and turn native, others said it was just a pretence, or the sunstroke had affected him, for the next day we were due to put to sea. What was to be done? George Kimmin spoke only when necessary. No one on the messdeck could get anything out of him. He just sat and stared, having a strange carved look on his face. It was his watch for leave, the last night ashore. No trouble there if he wanted to be absent from leave on the last run ashore.'

'Well,' continued Stripey, 'we consulted between ourselves, deciding that a few of his closest friends should have a quiet talk, surround him and instil some sense into him, point out his faithful sweetheart at home who was waiting, working hard and saving for him alone, remind him of his mother and father and their disappointment if he should do something foolish, get picked up with 80 days jankers and slung out of the Navy. So when the lads had finished supper and cleared from the mess for a smoke, they conveniently left George Kimmin to be interrogated and given advice. We started off with a friendly polite approach, saying we had been messmates all this time and we wanted to help him. We were concerned about the way he had been acting lately as if he might decide to abandon ship. To all this, George Kimmin remained dumb.'

Stripey looked up to impress as he continued, 'It was all to no avail. Not a bit of sorrow touched his heart, nor a shadow of regret passed across his face. "It is the Will of Allah" was all we could get from him.

'George caught the first liberty boat on the last day, carrying his parcel containing, we supposed, his white robes, and that was the last we saw of George Kimmin A.B. Those who followed him said the tent had gone from the sheltering palm tree, he walked straight past, as if knowing, keeping to the road that led to the open desert, and as they stood and watched his lone figure in the distance on tht dusty road, they did not trouble to go further themselves. They saw him disappear amongst a clump of shaded olive and palm trees. No George reappeared, but some ten minutes after, a robed Arab about the same height and build as George Kimmin as far as they could judge from a distance, stepped from the sheltered roadside, followed by a slender figure just like the silk-veiled Arab girl who had stood by the pitched tent. She was carrying a bundle for her Lord and Master, following in his path, taking up her rightful position in the true Eastern world. After we had put to sea, George Kimmin was reported overdue from leave with no trace of him on board. The Coxswain, with the Divisional Officer came to search his locker, finding all his private possessions intact. His bank book, with his last pay entered, had a note saying plainly and simply, "I give up all my Western ways, taking the Holy Faith of the East. It is the Will of Allah, signed George Kimmin."'

Stripey finished his story, shaking his head, adding to his listening audience. 'The more you see of life, the stranger it is.'

'Malta tomorrow. Arriving about 10.00 hours' was the information on the messdeck as we drank our 'Stand Easy' cup of tea. There should be heaps of mail waiting for us. 'I don't know if I'm still engaged to my London girl', said Jock to me as we sat together on the mess stool, 'I hope she has given me up.'

'Why,' I asked, 'did you go with her and become engaged? Losing your feeling so quickly for her is past my understanding.'

'Well,' said Jock, 'it was so convenient to have a place to go in London – saved my fare up to Scotland, and I never had to pay for weekends at her place. I have a wee Scotch Mollie up in the Highlands as pretty as the heather, a sweetheart from our schooldays. She is the only one for me', he said, slapping me on the back.

We ended up having a noisy argument.

CHAPTER SEVENTEEN

We sailed into Valletta harbour in Malta, with its terraces and pyramid houses bleached by the sun, dotted with the domes of many churches among the roofs of the clustered houses where the large, overcrowded Maltese families lived. However, the Grand Harbour was not for us. 'It's up the Creek for us,' Stripey had told us, 'where all the destroyers and smaller vessels are tied up away from the pomp and glory. Soon we shall be out to sea as escort for the big one's target practice, night firing, working up the flotilla into an efficient and workable force to take our place in fleet manoeuvres.

'You sprogs don't know what you've joined for. If you don't return to old England as fully trained seamen, you might as well throw yourself overboard.'

'When do we get leave to go ashore?' asked Ginger.

'You just don't,' Stripey answered. The old salt of many years' service was giving us the benefit and wisdom of his long experience. In his opinion we were not entitled to have the freedom and privileges which were denied him in his day.

'Well, I must go ashore,' I said to myself, as I looked from the guard rail of the *Ardent* across to the waterside where the moving buses and cars and horse-drawn carriages gave it a look of England in Queen Victoria's day. Some waterfront buildings had the look of a saloon about them, with large signs outside. We trained the gun round to look through the binoculars at it and read 'The Silver Slipper'. Further along was 'The Blue Lagoon'.

It was in the afternoon, when we broke off from washing the ship's paintwork, that the 'Killick' leading hand, detailed off as postman, came striding along the messdeck with a bulging bag on his back, giving out a deep throaty roar of 'mail arrived'. Instantly a lull descended on the company. Then a faint murmur rose, growing louder as men downed their cups and gathered round the Killick, with the bags of mail increasing as willing hands helped to haul them in.

The men called 'Here', 'Over there', 'This way' as the letters were handed out or flicked towards the answering calls. Now the murmuring died down and there was the sound of envelopes being torn open and almost everyone was reading news from home. The bosun's mate piped 'Carry on work', but no one took the slightest notice. At last my ears pricked up and caught my name and I pitched forward and caught my letter which I noticed straight away was in a blue envelope with small, even handwriting on it, which told me instantly who had sent it. I moved away to sit down on the nearest seat, opening my letter tenderly, as if I was touching the hand that wrote it, unfolding the thin blue sheets and catching the first few words in a quick glance. They read: 'My Dearest Darling.' 'Off the messdeck' ordered the P.O. as we sat or stood around reading our correspondence. He hustled us out and I quickly folded my letter up and put it away. Not until after tea in the quiet of the dog watches when most of the lads had gone ashore to sample Malta's entertainment did I unfold my sweetheart's letter and read of what she felt for me in our first few weeks of parting, the loneliness of not having me meet her on her nights off. There was an emptiness in her heart left by my departure which would never be filled until I returned. How was I standing up to sea life again? She told me to hurry up and write, not to do anything rash and to remember she was longing and waiting for my return. She sent 'heaps of love and kisses' and signed it 'your ever loving sweetheart, Isobel'. I did not go ashore that night.

We had been out in Malta some months. Our working time was almost completed. I had quite often been on runs ashore, visiting the main resorts of Hiema, Sliema, Floriana and the capital, Valletta. I was rowed ashore in the dyhuish and rode in the horse-drawn carriages, which reminded me of the faithful old carthorses at home, 'Blossom' and 'Beauty'.

There was much to entertain Jack ashore. The Navy authorities provided strait-laced entertainment, with tombola, and ping pong in the Navy Club as against the Maltese attractions and entertainment which included the bright neon-lighted cabarets and the notorious street of Stratio Strettar, which everyone in the Service called 'The Gut'.

One day we reached the quayside undecided what to do, for there were half a dozen from the fo'c's'le division, and it was recognized tradition that one kept to one's own part of the ship's company.

We took the bus to Valletta and somehow I drifted away from my party of shipmates to wander around on my own as I sometimes love to do. Anyway, I had a birthday present to buy which was a job for me alone, and I needed time to make my choice. I ambled along, eyeing displays of the many pretty and coloured presents that were on display in the biggest shops of the Island's capital. There were beads that sparkled, golden bracelets and earrings with rubies, but none of these seemed to serve the purpose. What could I buy my sweetheart, a long, long way off in England? Another year was to go before I must think about returning. I walked up one side of the street then down the other, finding nothing which I thought would be suitable. My thoughts turned more to her, and I wondered what she was doing just now. I had taken every opportunity to rush ashore by the first liberty boat to be with her some few months ago. Since being out here her letters had arrived each week, telling me all the exciting incidents that were happening in her world. Would she get tired of writing her weekly letters? There was such a long time to go. How can one stop having doubts and fears? I shrugged and gathered my thoughts together for lying in the shop window was the ideal present for her, the most suitable, worthwhile and useful thing, a tortoiseshell dressing-table set, so exquisite, just the right size and colour. The more I looked at it, the more I liked it. The feeling grew that my love would be overjoyed and most surprised when opening the parcel I could see her look of curiosity, and the puzzled frown in her wondering gaze as she took her time before opening it, then the smile of joy as she lifted the lid off the box and gently withdrew the mirror, comb and brush, that wrinkling of her brow, with a slow smile of pleasure as she placed them on her dressing table. Swiftly I walked into the shop, straightaway pointing out to the beaming shopkeeper the present I wished to purchase, adding that I wanted it to be sent to England. 'Yes, Yes,' he said, smiling more openly, adding 'for your sweetheart at home?'

'Yes, that is right,' I replied, 'I feel it is the most suitable present, the one I have been looking for and found without knowing exactly what it was.'

'It is lovely,' he said, spreading it carefully out before me, 'and would you like us to do everything, pack it and send it off from the shop?'

'Can you do all that?' I asked.

'Yes, Yes,' was the reply, 'all you have to do is sign these papers declaring on this form what is in the parcel, pay the postage in addition to the price and have no fear, everything will be all right.'

'Any duty to pay?' I asked.

He assured me there was none and showed me his book of many addresses in England that other sailors had sent presents to, just as I was doing, so I parted with my money and was given a receipt, and stepped out of the shop well and truly pleased with myself. With more spring in my feet I smartly walked up to the end of the street to the spacious building taken over as the Navy Club to help keep us on the straight and narrow path. There I settled down to enjoy a leisurely hour or two, until the gathering greyness descended rapidly, turning into the cooler blackness of night, when lights flashed on and more sailors came ashore, and life began to awake and stir from the drowsy, hot afternoon.

I rose to my feet and left the sheltered Navy Club. A nice walk all on my own would help to increase my appetite, and the starry hot night was something I really enjoyed.

I had been walking with head down, not noticing where I was going, when I noticed some steps which made me stop and look up. I became aware it had become quite dark and I had been walking by the glow of street lights. The steps led to some double doors with an illuminated arrow pointing to the word 'Entrance', and high above, sending out its message of invitation flashed in white brilliance against the dark night the name 'Silver Slipper'. I must venture in I thought, if only to talk to someone. I mounted the steps, pushed open the door and entered into a long hall with the usual tables and chairs with bar at the side, but in addition it boasted a stage with drawn velvet curtains. The time was too early for any show to be on and I was the first customer of the evening. I glanced around, taking my time in selecting my seat, as a waiter who had been leaning with his back to the wall came over. 'Yes, Sir,' he asked 'I want a drink which will make me feel hungry,' I ordered. 'I will get you just the right kind,' he answered. He hurried away to return with a bottle and glass, pouring out a nice red sparkling wine. 'What have you got me?' I asked as I paid him, 'This is an excellent drink.' 'It makes you very hungry for the ladies who will be coming in any moment, Sir.' 'You misunderstood me, I'm not hungry for ladies. I want a wine to give me an appetite to eat.' 'You are a very strange sailor, not like the others,' he replied.

I was thankful that I was alone, but now the waiter was talking to a lady who had just entered. One of the hostesses I supposed, but it was true that I had only recently arrived from England. Perhaps he was telling her that there was a young English youth to be baited and finally hooked, for she turned to come over to my table.

'You lonely?' she asked, in faltering English. 'Maybe I am.' I hooked a chair with my boot and nodded for her to take a seat. This I thought was a challenge to my ego, that I was just easy meat, to be preyed on just by her feminine presence. I didn't mind paying for her drink, for didn't I need someone of the opposite sex to unburden myself to? 'I only want your company for a little while. I am not buying you for the night.' I put my cards on the table and let go of her hand. 'I understand', she replied, 'it's all right. You have a sweetheart back home of whom you are very fond.' 'Yes,' I answered, somewhat taken aback to think this lady of the oldest profession could read me like a book. 'You're a lucky boy,' she told me, 'tell me about yourself, your home and country.'

When I had finished my tale, I said: 'You too have a family and home somewhere. What country do you belong to?'

'I am Hungarian.'

'That accounts for your pretty hair' – I took my hand from hers and stroked and brushed her long tresses.

'You speak nice words to me.' She shook her head and looked sad and lost as she told me of hunger and strife in the land of her birth which she could never go back to now she had left. When her mother died there was no work available, just a soup kitchen which only just kept them alive.

'I had to seek my fortune or starve. What can a poor girl do?' she asked, her voice rising. 'We are promised this, set to work, then given no money. My country was poor. All countries round here are poor. Only Britain is rich, so I find myself at Malta. Sailors have plenty of money and I have to live.'

'It's all wrong,' I told her, 'I am poor enough, but something may happen to make you rich. Try to reach America. That's where all the gold is. They mine gold in South Africa, ship it to America and bury it again. How burying gold makes a country wealthy I don't know. You try to reach America, where women have some freedom and status. In these countries of the Middle East you are suppressed and unrecognised. In England we adore and worship you. That's what makes us great. We place you high upon a pedestal and are forever trying to do gallant deeds in order to win a glancing favour.'

'I must go now,' I looked at the watch she wore, 'I have to meet my friends. I will come and see you again just to talk.' 'Please do,' she replied.

I arrived at a Maltese restaurant and ate a large meal served with prompt attention by the white-jacketed waiters. Their 'Thank you, Sir' as I rose to depart was a mark of first-class service.

I stepped out of their spotless restaurant, and turning deaf ears to the promise of a good time to be had, I reached the green and red curtains embroidered with letters reading 'The Red Garter'. I pushed them aside, stepped in and stopped.

Stripey and the boys were there in full force. 'Come on,' they shouted, 'You're late, have a drink?' A girl held a glass for me to drink from as she sprawled over a sailor's knee. I accepted a drink and squeezed myself into a chair. The atmosphere was thick with smoke, hot and stuffy, with a smell of overturned drink and heavy perfume from the ladies. Some couples were shuffling round the floor to the beat of the music, with the girls' loose revealing dresses swirling to show bare legs and when the music stopped the couples clung together, the men drawing the girls close, burying their faces in the partly bare bosoms, to make the ladies pretend to struggle and squeal in the few seconds the lights were switched out.

'Thank goodness,' I thought, 'there are not enough girls and I am late already.' The effect of the evening was already wearing the sailors down, as some were already in an abandoned state, and the efforts of their ladyfriends, for ever pressing their face into bare bosoms, were having little response, and as time wore on several seemed to fall into a semi-stupor, eyes closed, swaying on chairs, slowly falling over, muttering inaudible words, throwing their arms in hopelessness as their legs began to fail.

'Get me back to the ship, come on,' one shouted, 'I'm drowning', as he lay struggling on the floor.

'Pull him into the air,' said someone a bit more sober than the rest. I was only too willing to oblige, and with help, including his girlfriend, we hauled him out, pulling him by his bell bottoms.

Other sailors appeared from the dingy pit, held up or pushed along by shipmates and girlfriends. Stripey collected his wits, telling them all to pull themselves together.

The girls disappeared into the cellar and we struggled on in fits and starts.

'Keep an eye out for the patrol,' said someone.

'All those going off tonight meet down by the landing stage.' The word was passed along by some who were doing their best to keep some kind of order, and to lead their drunk and staggering chums on to a late bus. One or two who were too far gone were slung into a carriage, their mates holding them on the seat. The jolting of the horse-drawn carriage would soon bring them to their senses.

A small gathering of woebegone sailors now found themselves gathered together at the landing stage with the Maltese alongside, the boatman holding tightly on to the jetty rail as we stepped into his boat, some making it rock dangerously as they missed their footing and partly fell in.

The most delicate part of the evening was now to be handled, to keep everyone quiet as we got on board, for you never knew whether someone would start to sing just as we reached the side, or wanted to shout, telling the Coxswain just what he thought of him. Silently we swept along in the stillness. I looked round the boat, seeing Stripey had Ginger under his wing, for every time poor Ginger made the slightest stir Stripey pushed his head under the water, no doubt I suppose remembering the laughs at his expense caused by the wit of Ginger's tongue.

Now there was some cautious whispering, 'less noise', 'have a care'. We scrambled aboard, the quartermaster hurriedly saying 'Carry on', only too pleased to get us from aft to for'ard in case we roused the officers. We reached the security of the messdeck, but even then one had to be careful not to wake the messmates or those who had been on duty watch working late hours, who would not appreciate being awakened by a loud and amusing account of their messmates' amorous conquests ashore. In the morning there would be moans and groans and aching heads. Others would probably have to carry the burden until the age-old pipe 'Up Spirits', which had a most amazing effect of clearing thick heads and loosening the tongues of the sleepiest of sailors. Then they would all be anxious to give amazing accounts of this best-ever run ashore.

CHAPTER EIGHTEEN

I was becoming accustomed to destroyer life and hardened to the sea time. I had to muster to at almost every pipe, attend all the upper deck work, help to clean and paint the whole ship from the top of the mast down to the water's edge, even scraping the ship's bottom when 48 hours could be spared us to go in dry dock. I was forever mustering and was excused nothing. But one day the Gunner's Mate posted up a quarterly list of changes, and behold I was informed by my messmates that I had the cushiest job of any: wardroom sweeper, with quite a number of duties being excused as a result. Every morning, while the other sailors were scrubbing the deck, I would carry on aft to clean the wardroom. I could now prepare the food for a dozen hands and decide on the menu for the day.

Daily I would scrub and polish, giving much care and attention to the cleanliness of the wardroom, where the officers dined and the King was toasted sitting down. (This custom dates back from George III, when His Majesty, rising to acknowledge the toast, had bumped his crown on the bulkhead.)

In the late forenoon, Jimmy the One, after pacing the quarter deck, seeing that everyone was pretending to be busy about ship, would descend to the wardroom, sprawl in a deep-cushioned chair, ring the bell and order a 'Horse's Neck'. I would be dusting the paintwork or polishing the brass round the portholes as Jimmy leisurely drank his cocktail. I decided when I got back home to go into a hotel lounge, drop into a divan and order a Horse's Neck, just like a naval officer.

After some months the whole Mediterranean fleet began to do manoeuvres on a war footing. Every ship cleared away for action. Boats were swung in and hooked down, for if anyone fell overboard or was swept away as we made a 90° turn, it was 'swim for Malta or go under', as Stripey informed us. I had closed each porthole of the wardroom, the deadlights down and screwed up tight with all watertight doors and hatchways fastened down and closed, even those on our upper messdeck. 'What are we doing for eats?' Stripey asked Ginger. 'Food in wartime? You have no time to eat!' said Stripey. 'A hungry man is a fighting man. Filling your belly only makes you fall off to sleep, and you know what the penalty is, you bottle-fed sons of landlocked mothers,' he bawled.

'What is the penalty, Stripes? Are we to be tied to the gun and so many strokes of the birch delivered by the Gunner's Mate, with you standing by to rub salt into the open wounds?'

'"Birched"?' remarked Stripey, looking down on us with a look of scorn on his weathered face, 'you don't get off that lightly in these days of quick decision and swift action. One lapse of duty and the whole ship and crew can go in a flash.'

'Well, what then?' asked someone.

'You are hauled up on the bridge', continued Stripey, 'before the Captain, the Coxswain stating the charge as there is no time for formalities in wartime. Remember the Captain is King, judge and jury. You might see him lift his hand from his coat pocket because that will be all. The others close by will hear a sharp report, then the rating will slump in a heap, a red gush of blood oozing from a head wound which the revolver bullet will leave behind. That's the penalty you pay in this age of modern warfare and don't you think the cause justifies the penalty?'

The younger ones hesitated, staring at the speaker, realising for the first time there must be a certain amount of truth in his statement. Before we could address further questions to Stripey the shrill note of the Bosun's pipe with the call 'Hands to Action Stations' was heard. There was an instant rush as men scrambled to their appointed duties. I felt the deck vibrating under my feet as she gathered speed, settling her stern down in the water, the white ensign flapping, strung out to its full extent from the mastheads. The men were alert and ready to deal with the new menace of attack from the air. The duel between gun and aircraft was yet to be fought out. The surrounding force of destroyers with sloping funnels lay further back as their bows lifted up by speed cut through the water.

As I stood, looking out to sea, impressed by the power of the fleet, I remembered the bible story of David and Goliath, how Goliath with all his protective armour, possessing the most deadly of weapons, which could topple David's head from his shoulders in one slight stroke, had been slain by a simple sling that hurled a stone, guided by the unerring aim of David's arm – a skill obtained in his everyday life as a shepherd boy.

Were there not the lurking U-boats, hiding below the surface, ready to strike an underwater blow; or a mine anchored in our path, besides aircraft suddenly diving from overhead; or over the wave tops was there something simple or ordinary able to halt this fighting force? My enquiring mind was always ready to sort out the possibilities of a flaw or weakness in a build-up of strength, just like David's plain and simple sling.

'Alarm, alarm,' rang out. 'All guns load with barrage, aircraft approaching from astern.' Instantly there was pandemonium all around as guns swung round. The ship began to zig-zag as we loaded numbers on our gun, hurried to set the nose fuse correctly on the anti-aircraft shells. 'Be quick and set that fuse you –' roared the Captain of the gun. 'Slam that damn shell home', as it was snatched from the fuse-setter's arms by the infuriated Leading Hand, closing the interceptor with a quick snap that would let the control know by the tell-tale light that 'A' gun was ready to fire. They were already calling over the headphones enquiring what was delaying 'A' gun, but no sooner had the gun been brought ready to fire than the ship turned to port. The trainer struggled to train his gun to starboard as engines whined over us and an aircraft dropped bombs of flour bags, one of which burst, scattering over the bridge, dusting the Captain and his staff with specks of white dust, thus making them casualties, also causing severe damage to the wheelhouse and mortally wounding the helmsman. With the bridge and wheelhouse now out of action, steering would be controlled from aft, but while all this was happening the Gunner's Mate had rushed up to 'A' gun, demanding to know why the gun had been the last to come to the ready. 'The fuse-setter, where is he? Come here, what the hell took you so long? You have had all this practice to set fuses and now you let everyone down.'

'Couldn't help it,' he replied.

'You idiot, why not?' snapped the Gunner's Mate.

'The fuse key dropped overboard when the ship slewed round,' was his reply.

'It's a pity you didn't go with it.' The words of anger were cut off as a shout of command came from a Midshipman running up.

'A gun's crew "out collision mat" beneath the bridge on the starboard side.' A hole had been blown in the ship's side on the waterline. The Gunner's Mate reported to the 1st Lieutenant, who was now in command. The steering was being operated from the aft while the for'ard repair party, under the Engineer, rushed around shoring up, working to restore the damaged wheelhouse.

'Where is the collision mat stowed?' enquired one of the 'A' gun crew.

'Astern of the bridge. Follow me,' shouted our Leading Hand, rushing off, only too pleased to get away from the wrath of the Gunner's Mate and to do something to show off his knowledge of seamanship, as we quickly untied the mat from its position, shouldering it and marching off under his direction to the starboard side under the bridge, care being given to it being unfolded correctly with the for'ard and aft guides being run out. The lowering line was well backed up with a block and tackle to take the strain, the bottom line having been taken for'ard round the bows and lowered into the water so as to haul from the port side on the bottom line.

Up rushed the Buffer, not to be outdone by a Leading Hand.

'Which side of the mat is to the ship side?' he asked.

'The thumbed side,' was the reply.

'That's right, lower away. Stop. Make fast. Take up slack. Send a messenger to the 1st Lieutenant. Collision mat in position and secured.' He was taking charge from the Leading Hand, now that the operation had been almost completed.

'Right, back to your gun,' was the order given with a wave of his hand, for now the steering had been restored for'ard, the battered wheelhouse and bridge was in normal condition, with the Captain restored to life and in command.

The *Ardent* had not fallen out of line and was steaming majestically along. The Captain hoped the Admiral had observed the swiftness of the collision mat being placed in position, thus gaining a favourable report for his crew's ability. We remained huddled around the gun after our hectic half-hour's excitement of repelling aircraft and getting damaged. Around midday one hand was allowed to go down to the messdeck to bring back a bag meal of sandwiches and rolls for each member.

Repeated dummy runs of attack took place to test our alertness during the day, with communication being tested every twenty minutes. No time was allowed for relaxation or sleep, although a welcome cup of ship's cocoa was brought round near midnight. Well into the night we huddled up half asleep, well wrapped-up to keep warm in our coats and oilskins, trying to fit ourselves into the small gunshield for protection from the spray and wind. Then from the bridge we heard the look-out report, 'Two Very lights on the port beam'.

Our Leading Hand gave a mighty heave, sending us almost flying as we tumbled from the shield, bringing the gun to the ready. The ship sprang to life, turning to port where the Very lights had been fired. Cordite fume filled my nostrils, and star shells burst high in the night sky.

'There she is,' someone shouted as the light from the fading star lit up the stern of the attacking ship. Now the enemy had been spotted and we were in hot pursuit. Other star shells now dropped over the target, with the remaining guns firing at the lit-up vessel as she twisted and turned, trying to escape into the darkness.

Soon, now, dawn would be breaking. This is a vital time in a war operation, for who could tell whether the first flush of light would reveal an enemy force that would get in the first salvo? Britain's silent Navy must never be caught napping.

When we were able to relax and have a chat, Ginger said: 'I expect the Home Fleet have slipped by in the night and have dropped anchor in Malta while we are out here searching. Just think of them going ashore tonight, waltzing round the Blue Dolphin.' Then he went on to a favourite topic.

'Look at Sandy here (referring to me), still writing home regular, but how much longer will it last? A sailor's life should be gay and free, drifting from port to port, enjoying life and free with his money. Love, leave them and be gone.'

'I don't agree,' I said, piping up when I suppose I should have kept quiet, 'isn't that being greedy, taking all from life and giving nothing in return, living for today and forgetting tomorrow? I have someone to care for, someone back home, someone to write home to.'

'You're crazy,' they replied, 'I bet she is having a good time with a big fat Marine.'

Why did I listen, to let their tormenting and hurtful words fill my heart with remorse and misery?.

Perhaps it was this glorious morning, the vast, glistening empty space, casting its spell of loneliness, gripping me with tender thoughts of those so far away. I walked a few yards. I couldn't help feeling a little homesickness, a longing for her. I felt empty at not being able to share this wonderful morning. My silence was enough to make the others aware of my discomfort.

'Don't take it to heart,' they called out, 'keep your chin up. She may be the one that will wait', then adding that thousands of others don't, and smiling and winking at each other behind my back.

The day wore on, but even with war routine we discovered there was a certain amount of work to do. Soon the rumour spread that we had turned for home base and we had to take in tow another destroyer which was supposed to have been damaged. This took some of the attention from the Gunnery side. We drew near enough alongside to throw over a line which was quickly tied to a rope, then by careful manoeuvring whilst we hauled away, the other destroyer paid out her ropes, wires and cables, much care being taken that we did not foul our propellor. Jimmy the One gave his personal attention as he anxiously looked overboard to see the sagging wire dipping in the water, then loudly gave the order to haul away until at last the cable was secured aft and the go-ahead was given to proceed slowly ahead, with our sister ship being towed back to Malta to enter harbour again after almost a week of strenuous sea time and experiencing the hazards of what might befall us in war.

Time wore on, and Malta was becoming more of a home base to most of the ship's company than our actual home port in England. Out here nearly everyone went ashore. There were those who slipped off at almost every opportunity, who had become attached to a certain saloon.

The married ones could only afford to go occasionally, because their allotments home and their devotion to their family left them very little money or desire to revel in the debauchery of the seamier side of life. There were the organised parties that were laid on when the ship made a courtesy visit to an island or large port of the many smaller countries that surround the Mediterranean sea. There were sporting trips and time given to train whenever possible and of course, swimming. Hours have I spent swimming in the clear warm blue water, or lying on the sandy shore soaking up the sunshine.

But now this orderly life of ours was about to be shaken by an event that had its repercussions in all the capitals of the great countries. Britain who had been so complacent had signed the Armistice with relief after a stalemate on the muddy battlefields of the Continent. The German fleet had surrendered and lay at the bottom of Scapa Flow, a watery grave for a nation who dared to challenge our sea power.

But what did the majority of people gain from the War? It was bread for hungry mouths and better conditions they wanted. Why should a few own the land, factories, even their homesteads? The only thing they didn't own was the road running by, but they taxed you all for its upkeep. This was the bitter cry wrung from the hearts of the poor people against the glitter of wealth from the rich.

One side had to give way, for the Liberals, the middle party, who might in time have gradually uplifted Britain's labour class, were now crushed to a pulp betwween the two powerful parties.

England fought her revolution and the workers gained their crust of bread and the ruling classes released their grip of power, this rule which had been inherited and bred in their families for several generations. Was it altogether wise that men with bitterness in their hearts against one class should rise in a short time as world statesmen with an enormous Empire to maintain and world strategic force of armament to control? Thank goodness the men of aristocratic breeding retained their composure, their nobility, their gracefulness although badly shaken. Their leadership and understanding of the world's affairs was still needed by Britain in its hour of peril, which was still to come.

What was happening? Something mysterious. The Chief Stoker had hurriedly come on to the messdeck and shouted down the hatch to the stokers' mess for certain hands required immediately, and Stripey had seen the signalman rush down the ladder to the Officer of the day who had gone straight to the Captain's cabin.

I returned to the Mess, for everyone had stopped working and seemed intent on listening to pick up any thread of news that would throw some light on what it was all about.

'We won't know until the Captain returns. He has been called over to the *Codrington*, where the motor boat is laying off waiting to bring him back.' I gave this information to the others, who looked on silently, then turned their gaze to Stripes as if it was up to him to let us know with all the wisdom of his long experience, and we paid our respectful attention to what he had to say, but even Stripes seemed as if he had no clue until someone pumped him, asking him directly what he thought was happening.

'Are we going to War, Stripes?'

'War?' he said, 'There soon will be one if we allow our dwindling forces to get much lower. Reduce, economise and feed the starving sounds all right in the papers and on the radio.'

'How do you expect people to fight for their country if they are not fed?' spoke up a Liverpool lad.

'I know there is something wrong', voiced Stripes, 'which needs putting right, but whilst they are arguing over the affairs of the country they are neglecting the affairs of the world. They can see no farther than the industrialised cities or understand their hardships. As soon as they get a little money in their pockets you'll see, they will be like the rich, probably worse. Look around. Haven't you seen far worse conditions out here than ever existed in our own country?'

'Whilst we go on cutting down our forces other countries will build theirs up. I know these foreigners only too well. This Hitler and Mussolini. These arrogant Germans are not finished yet. They still think they are the master race, and these Italians, don't ever turn your back to them. These soapbox politicians that have now sprung up in our old country are playing into the hand of the dictators and it won't be many years before they call the tune. Maybe that is what they are doing now, deploying their forces for another showdown in the future. To lay off is a sign of weakness. A little hunger is better than overfeeding. Hungry eyes are cast upon this Empire of ours, and evil minds are puzzled how a small country can hold it, but the moat which flows round our island is guarded by sea power. To get at us they have either got to burrow underneath or fly over the top, and it is most likely the air that we have to fear; if these soapbox politicians undermine our constitution by neglecting to keep up our forces, it will mean another Great War.'

Now everyone had stopped talking to listen. The shrill note of the Bosun's pipe could be heard. Ears were strained to catch the message that followed the warning whistle. 'All leave cancelled; the ship is at two hours' notice. Both watches of hands will fall in in ten minutes' time.'

We all stopped talking as if to let the effect of the message sink home until Nobby Clark broke the silence by declaring: 'These damn Dictators have robbed me of my run ashore. "Join the Navy and see the world" – and as soon as you are out in the world all leave is stopped.'

The Senior P.O. called us to attention, smartly saluting and reporting 'Both watches of hands all present Sir.' Jimmy acknowledged leisurely, in navy officer's style, touching his hat and saying quietly 'Stand them at ease.' He looked down, and then stepped on to where the deck was raised slightly higher, circling his arms as he said, 'Gather round', for the outlet of air from the engine room made listening rather difficult.

'I expect you are all wondering why the sudden state of preparedness that is obvious to everyone on board is taking place. All I can tell you is that the international situation is somewhat strained at the present, and of course, being on foreign service the Mediterranean fleet is ready to fulfil its duty if called upon at very short notice.' Someone in our midst called out 'Hear, Hear', followed less faintly by another. 'Silence', snapped the P.O., but someone must have trodden on the foot of the enthusiast for his 'Hear, Hear' was suddenly changed to 'Oh, oh you awkward –'. 'The first casualty of the war', whispered Ginger.

'Quiet' was whispered from all around as Jimmy continued. 'Any information that can be released will be passed on. In the meantime we must get ready to meet any emergency. That is all.' He nodded his head for us to fall back into our usual part of ship's muster, with the everyday routine being carried out, only now the fo'c's'le men would be securing the ship's wire to the buoy, with parts of the ship securing for sea. This only served to make us more tense. The Gunner's Mate was looking much more serious than usual, giving earnest attention to his party, with the Gunnery Officer in attendance.

Already the torpedo branch were unscrewing the practice heads of the torpedoes to replace them with large, black warheads. I picked my way past them, looking at the weapons that could deliver an underwater blow, feeling a little bit excited to think that I was in the midst of so much war activity. This was real. There would be live ammunition in the breeches of the guns, if we did go to sea.

There was much activity going on on the upper deck. I heard the motorboat come alongside the gangway, the Captain piped aboard. Some time afterwards, I heard the Officer of the day giving orders to the Coxswain of the motorboat to lay off, to be ready for hoisting inboard. There was no mistaking now. It was definitely out to sea. All hands were being piped to hoist in boats, followed by change into rig of the day for leaving harbour. I would join them, if only to bid farewell to Malta. I made my way forward to the messdeck to join the others.

Gallantly we sailed out of Malta with the flagship leading our flotilla. Tiny Malta began to fade away. We were dismissed and piped to close up for action stations in five minutes' time. In the few spare minutes we had to change in, many voices aired their opinions of what they thought might be happening, which included an attack on Mussolini's Italian fleet, and a raid on the Kiel Canal. Suggestions were made that we were about to bottle up the Mediterranean by stationing half the fleet at Gibraltar and half at the Suez Canal. One hopeful voice remarked we might be going home to get installed a secret weapon while the home fleet took over our duties. This secret weapon was a deadly ray, which could stop aeroplanes in flight and send them crashing down into the sea. It could even stop enemy battleships when the ray was beamed on to them.

For me it was a memorable occasion as I watched the whole fleet in squadrons and flotillas manoeuvring to take up their position in line, ploughing through the sea. 'Oh, England,' I spoke inside me as pride swept and held me spellbound at the sight of this vast fleet of my nation's power in full operation. Before me lay my country's history. I saw the pictures of time from King Alfred building his boats on to Britain's famous admirals through the ages, Raleigh, Drake, Hood and Nelson, the greatest of all, into the modern age of Beatty and Jellicoe. Perhaps this very moment would make history as I stood watching this spectacular happening.

The whole Fleet was now out of sight of land, altering course and steaming for some destination undisclosed to us as yet. We would now settle down to war routine at sea with the fleet manoeuvring with the battleships in line, a squadron of cruisers as a covering force, and another squadron steamed over the horizon to await a favourable opportunity to do a dummy attack, whilst we on our little destroyers dashed around giving protection from the lurking submarines or torpedo attacks on the cruisers, or went out as scouts to investigate anything suspicious, sending back and receiving messages.

It was three days later when it was piped round the ship that the fleet would anchor in the morning in the harbour of Alexandria, the large seaport of Egypt. The diplomacy behind this show of strength was the Italian invasion of Abyssinia, the first move of Mussolini, the Italian Dictator, to expand his Empire. It did not stop the Italians from shipping their troops through the Suez Canal, to overrun this desert and rocky country by weight of numbers and modern equipment against tribesmen on horses and in some instances still using the spear. A sure case of a bully using his weight and strength to crush the weak and defenceless, but it was not for me to understand the moves behind the white gloves of our statesmen. The sanctions which were imposed under the League of Nations were only laughed at and led to more boasting by the strutting, domineering Dictator, a sure sign of stimulating his desire to further his ambition and advances in the future. Didn't it make him a hero in the eyes of his country – one who dared to act and get away with it against the powerful but hesitant British?

Alexandria was a harbour of sunshine, but not for long were the ship's company of H.M.S. *Ardent* to rest and bathe in the warm clear water. With two other destroyers we were despatched to spend our time anchored off Suez, keeping an eye on the Italian troopships passing through, with an occasional visit to Cyprus, Haifa and back to Alexandria.

It was during this time that I lost my job as wardroom sweeper, where I spent much time in the small pantry with two Maltese stewards, tasting the officers' delicious pies and cakes given to me from time to time as I waited for the 1st Lieutenant to remove his sprawling body from its relaxed position on the wardroom's cushioned chair, having to face up to my messmates' sniggering remarks that I was snivelling and mixing with the lowly Maltese. But now the Gunner's Mate called me by my full name, very seriously, as his job had made him that way through dealing with officers and men and feeling the responsibility of his rise to power. I stopped with concern and wonder spreading through me.

'Yes, P.O.' I replied, thinking fast. Had I done anything wrong? He rustled his papers with a list of names, looking down still with such a serious look on his face that I began to look at it too, edging round to get a better view.

'Oh here it is. Now you', he spoke, 'wardroom sweeper going as Quarter Master, take over duty at noon, the 1st of the month.'

'Quarter Master,' I gasped, going back on my heels, not wanting to edge nearer. I took two paces back and if it hadn't been for the guard rail, I would have stepped in the drink.

I opened my mouth, muttered that word 'Quarter Master?' and closed it again. I felt worried and concerned as I watched him put a big tick beside my name and then turned away, wanting to know where he would find A.B. Smith. I wasn't able to tell him owing to shock and he hurried off, not listening or waiting for an answer. Slowly recovering from the shock I walked steadily along the deck to my mess, sitting down and thinking deeply. I would now be a watch keeper, would move into another mess to a deck below with the miscellaneous branches, but what kept me quiet was wondering whether I would be able to do the job. At sea the Quarter Master supervises the steering of the ship, and in harbour he keeps watch under the Officer of the Day and is responsible for seeing that the whole ship's routine is carried out.

Soon now the whole ship's company would know, and I could already feel the looks being cast my way, the nods and knowing winks as I turned my head, the grins, I could hear them saying, 'Well, that's shaken him. He's had an easy number. Is he going to be able to carry it out?' I felt as if the whole messdeck's eyes were on me.

I went about my duties giving much thought and consideration to my new job, thinking to myself 'Why did the Gunner's Mate put my name down for this particular number when there were others, young and more capable of responsibility than myself?' I never pushed myself forward but always seemed satisfied to run along with the back numbers. Did he see a shining hope in my ability, or did I need a shaking up? After going over all the details in my mind and sorting out the reasons as to what was best for myself and what course to take, I decided I must accept this challenge to show all my doubtful messmates that I was capable of carrying out these duties. I did feel inwardly that I had something stored away which I could reveal only in a crisis such as this. I must not let myself be humiliated by saying I could not do it. My pride would not let me.

With my mind made up I immediately brought out my seamanship manual, swotting up on all the duties a Quarter Master has to perform. Then there was the rating from whom I would take over. He was going home to England and was only too willing to give me all the help he could. He showed me how to write up the log, as everything of importance must be logged, even to the force of wind and sea running at the time of signing off after one's spell of duty. 'Stripes,' said I, waiting in the mess to catch him alone, 'You know this new job I am about to take over, well I have some misgivings as to my ability. I don't think I have enough experience as yet to do the job. I am not at all sure of myself.' I spoke quietly and earnestly as one does when confiding in another. 'Why the Gunner's picked on me I don't know.'

'Why you lack confidence in yourself I can't understand,' he replied, looking at me in a 'pull yourself together' way. 'So far you have done all right, haven't you?' He had a searching look in his eyes.

'Well, it has been a struggle,' I remarked.

'You want to have more confidence in yourself. Don't let others break you down. Just think, they might be a bit jealous of your selection, and their assurance of their own capabilities may be a way of concealing their inability, so you just go ahead and do the job like a real seaman or do you want to remain like me?'

'Like you, Stripes? I should be proud to be like you.' I was taken aback. 'You think I can do this job?'

'Why not?' he answered me. 'You are as capable as any of the others.'

'There's one thing troubling me. I don't think I'll be able to blow that Bosun's call properly. It's worrying me a bit. I never could whistle a tune and how I'll ever learn to blow a note on that I don't know.'

'It will come with practice,' he assured me. 'Get hold of Godfrey the officers' servant. He has got time. Go down the after tiller flat, thats the only place on board where you can practise so as not to disturb others.'

I thanked Stripes and went to make arrangements to carry out the advice given, feeling a little more confidence in myself after our talk together.

I went down to the tiller flat with old Godfrey A.B. teaching me to blow a note. We called him 'Old Godfrey' not because he was old, but because he had a bald patch which made him look more than his age. 'Blow! Close that hand, now open,' he instructed me. 'That's what controls the little pea in the hole.'

I blew until I was red in the face and clumsily opened and closed my awkward hand, doing everything wrong and cursing myself for not being born with more of an artist's touch. How could a clumsy clodhopper who had been brought up to handle the muck fork be taught to have a delicate touch on a tender instrument which sounded beautiful thrilling notes?

The day came when at last I had to take up my bed and walk: a mess change. I packed my bag, removed all my personal gear, shouldered my hammock and departed to the messdeck below. It always reminded me of moving from one village to settle in another, when after a few days of joining the new community one became accustomed to their habits. No more dirty overalls, but a clean uniform with white shoes and white 'dicky flannel' and to denote my authority the silver chain that one placed under the collar, hanging down in loops similar to a lanyard which had attached to it the Bosun's call, so that at an instant one could alarm the whole vicinity. Taking his place by my side was a young Ordinary Seaman as Bosun's Mate. He was my messenger, carrying out duties under my instruction, and I myself was answerable to the Officer of the Day. I must say I was conscious of my job as I signed the log and commenced my first spell of duty as Quarter Master, pacing up and down the Quarter Deck by the gangway, with the Bosun's Mate by my side just like the Captain, pacing up and down taking his evening's stroll, calling his No.1 to keep him company.

As yet, I kept very close to the gangway, instructing my Bosun's Mate to carry out the routine of calling the hands to muster. He had carried out these duties under the old Quarter Master. I performed my duties smartly, saluting the Officer of the Day and answering 'Aye aye, Sir' in my best nautical manner. When I did call the ship's company to attention it was with a full deep voice, powerful, distinct and commanding. I really surprised myself bellowing out in front of my old messmates, trying to show not the slightest sign of confusion as I patiently bided my time.

As yet, I hadn't fulfilled my duty. I still had to prove myself. Coming to from my faraway thoughts and turning to the Bosun's Mate to check the time and go the rounds to see all was well. Tomorrow we were proceeding to sea and then I would be taking the helm. I felt elated and yet dubious. I did want to be a success, to show my messmates that I could do the job. I felt a lot depended on me, this very day as I signed the log, ending my middle watch to get two hours sleep before the wakening of the ship's company.

'Relief Quarter Master to take the wheel Sir.' 'Very good' was the reply coming down through the voice pipe from the bridge above. 'Here you are, course 040, both engines half speed ahead, revolution 140', turning his duties over to me as he stepped aside. I grasped the spokes of the wheel, taking up a standing position where I gazed into the azimuth mirror to see the degree on the compass card in minute movements pass the lubber point which denotes the ship's head. 'Steady her', he whispered, 'by putting opposite helm on.' He assisted me by turning the wheel slightly as the lubber point began to move. 'Always remember the lubber point is the ship's head and the compass card does not move. Ease her. That's right.' He gave me encouragement as I turned the wheel or eased it, keeping on course now. From the voice pipe came the order 'Starboard 20, Starboard 20, I repeat, East to ten, East to ten, 10 of Starboard on, wheel amidships, wheel amidships, Port 10, Port 10, 10 of Port on Steady, Steady 146'. I called up through the voice pipe and repeated again when the ship was steady 'On course'. I felt the ship swing round; for me at this moment it was a great ship. I took my eye from the compass, seeing the bow swinging round, and applied opposite helm to check and steady her swing. I dared to take my eye from the compass only for a second. I was steering and the ship was answering, swinging round slowly as opposite rudder checked her. I held her steady and called out the degrees in three numbers, reading from the compass card the exact degree which the lubber point was on, then I got her back on that course by applying a little opposite helm and easing off the wheel.

More orders are given as the watch proceeds. I feel more sure of myself. The tenseness wears off and I become more relaxed. Before the watch is over, I can let the wheel spin back and the faltering is gone from my voice as I repeat back the orders in a clear tone. I turn over my watch to my relief – a little reluctantly, thinking it has finished so quickly.

I step out of the wheelhouse with my head held high, down the ladder from the bridge to my mess. My face must be full of confidence – I feel happy and great. What must I do next but take pen and paper from my ditty box to write home to my love, and tell her that I am now a Quarter Master. In harbour I see the ship's routine is carried out and at sea I steer our great big ship.

A shadow stills hangs over me. It is that cursed Bosun's Call, but with renewed confidence from carrying out the most difficult duties, I throw myself wholeheartedly into picking this up. As soon as we are in harbour again I am down the tiller flat, carefully watching my instructor as I follow his every movement, blowing whilst opening and closing my hand, and then suddenly it comes – so easily I wonder why it ever took me so long. I blow and blow till I am satisfied with myself, even to that little twiddly bit at the end of the pipe down. Until now I had given a straight blast when I piped the orders and had had to put up with my mates' sarcastic remarks of 'Learn to twist that tongue'. 'You ought to be able to waggle that enough, by the way you use it to wriggle out of work.' I'll quieten their remarks tonight, I thought, by giving them a few more blasts under their hammocks until they shout for me to go away.

I signed on my forenoon watch, all hands were busy working about ship. The officers were in their cabins, writing reports. I scanned the harbour and saw making towards our ship a motor boat, with pennants flying, indicating that foreign officers were about to pay a courtesy call on our Captain. In an instant I had summoned the Officer of the Day, who quickly informed the Captain. I took up my position at the head of the companionway with my Bosun's Mate by my side. The Officer of the Day buckled on his dragging sword quickly, stepped on deck and our Captain rushed up from his cabin below, adjusting his gold laurel-leafed cap.

As the foreign officers stepped on to the gangway, I in true traditional style, with my Bosun's Mate taking his time from me, piped them aboard. The Officer of the Day, in all his finery, raised his right hand in salute, his left hand holding his sword. All around stood rigidly to attention, then advanced to meet and introduce the Captain. After a word and a handshake, the Captain disappeared with the officers following, down the hatchway to his spacious cabin.

I turned smartly, piped the 'Carry on', and felt I had grown two inches as I felt a surge of confidence overcome my doubtful fears.

During these long months of activity, the fleet was based at Alexandria. Much of our time was taken up with patrolling from Alex to Haifa and Port Said, with an occasional visit to Cyprus. Haifa, the seaport town of Palestine, built under Mount Carmel, had a sheltered harbour, where destroyers could tie up stern to the sea wall.

'This is the up and coming country', Stripes told us, 'Where the wandering homeless Jews will return to their promised land, which one day will be their home and country and will rise into a great nation.' At that time it was mandated to Great Britain, having been captured from the Turks in World War One. Britain had promised to set up a national home for the Jews, but as the Arabs were averse to the Jewish immigrants riots and bloodshed had developed, and Britain found it difficult to find a solution to this complicated problem.

I was impressed by this land where the priests, from their towers, called the faithful to prayer, in wailing tones. Wherever they were, the Arabs dropped to their knees, facing East, and bowed in prayer to their God, the Holy Allah.

From the bridge of our ship I could see the terraces of white houses, stretching in lines across the lower slopes of the holy mountain. On the opposite side I could see the sandy desert lying flat and golden under a blue sky. I could see long desert caravans, crossing in single file with their shapely outlines showing clearly against the yellow sand. I could see the laden beasts moving sedately over the skyline into the vast emptiness of the horizon.

I went ashore and mingled with the people in narrow streets, the bazaars where the native workmen carried on their ancient crafts. I saw wonderful designs of angels and the Son of God, pictures that reminded me of the Old Testament, and of the faith of people who turned from stone images and began to worship God. I was back a thousand years into biblical times, walking up the terraced steps of the Holy Land. Beggars grovelled, hands tugging at the hem of my garment. I felt desperately sorry for these deformed, blinded, pitiful figures, knowing how little I could do for them as I put a small coin in a grasping outstretched hand.

As I walked back to the ship from this shore visit, I let my mind wander back to my infant days at school when I was taught my first prayer. I remember learning the Scriptures and Commandments. 'Thou shalt not steal' was the 8th Commandment, and I had been caught stealing rosy red apples from an orchard. How ashamed I had been when my father informed me that I had been seen throwing sticks to knock the apples down. Mother had scolded me, saying that Jesus wouldn't like little boys who broke the 8th Commandment. I had worried for a whole week that now I wouldn't go to heaven. Finally, I asked before I dropped off to sleep one night that if I became a man I could travel to that faraway land where Jesus lived to ask him to forgive me so that I would still be admitted to heaven. Here I was, right now, walking in the Holy Land. In a few days' time I would be going on a day's outing to Jerusalem.

I visited the Holy City of Jerusalem with a party from the ship, travelling in a ramshackle bus along the winding road, through the almost barren land of rocky ridges. I saw attempts at cultivation in age-old style, with oxen pulling wooden ploughs behind them, with a stately Arab following, so slowly, as if he had his whole life in which to do it and there was no hurry. The night would come, dawn would break forth into another day. One would follow the other in this land where one felt time was born. In this ancient Holy Land, wise men of learning have studied the heavenly bodies and written the gospels. The modern world had been given the wisdom of the old. I felt as we rushed by in the bus, seeing life as it was in the old world, that we were rushing to our doom. The age of modern machinery would burn itself out. In centuries to come we would set fire to the world, then a calmness would descend until life began again, slowly, as I was seeing it now.

These thoughts filled my mind as I journeyed into Jerusalem. With a feeling of awe I realised I was now standing in the Holy City in the very heart of the land where my religion had started. In Bethlehem I stood at the very spot where the bright star stopped above the stable and the wise men brought gifts to the holy infant. I was shown the cattle shed where Jesus was born, the manger in which he was laid. I saw the road from Bethlehem on which he rode on the ass to enter Jerusalem; the wailing wall, with Jews wailing in mourning for their lost city, praying that one day their homeland would be returned to them; and finally the magnificent church built where Jesus spent his last night praying in the Garden of Gethsemane. The gaunt old tree trunks which were 2,000 years old, so the guide informed us, and were still flourishing. Branches grew from these withered trunks, twisted, worn and bent, as though they had lived agonising years throughout the centuries, waiting for the day of judgement when the old shall be made young. I gathered some of the small, fading leaves to send home, a faded leaf from Old Jerusalem, to be pressed in my album of memories.

Months went by, during which Mussolini transported his troops into Abyssinia, and the King of that country took refuge in England.

The Fleet was still based in Alexandria when the King of Egypt died and his son was recalled to be crowned King. The whole of the Mediterranean Fleet turned out to welcome him. There was a double line of British warships stretching far out to sea from the harbour entrance with sailors lining the guard rails, cheering him as he passed through.

The ship's company, having accustomed itself to foreign service, now regarded themselves as old stagers. We had reached the halfway stage of our commission. As a Quarter Master I had become a responsible member of the crew. I felt the time had come to try to improve my position.

Letters from my sweetheart were still arriving regularly. It had been well over a year since I last saw her, but I had wondered about the long, long months, with just a promise between each other. She was employed as a Cook-General at a Doctor's house, and I was an Able Seaman in His Majesty's Navy. Our joint income was, at the most, 30s. per week. We were 2,000 miles apart and likely to be apart from each other for as long as two or three years. It had to be the pure golden threads of true love, to withstand the longing that must be within her to be free to enjoy dates and parties instead of quietly sitting down and writing to her sailor. I wondered sometimes what I had to offer as I sat thinking and answering her weekly letters. The least I could do was to try to get a step further up the ladder. With this in mind I put my name down to attend school and have instruction for selection for the course for Leading Seaman.

Our Navy was very selective, even on the lower deck, so in my off watches I sat with pen and paper doing sums of the lowest common multiple, swatting up on my seamanship and being taught to drill a squad of ratings when they could be spared if we had the good fortune to be tied up alongside the jetty.

I stuck patiently to my studying and boldly gave command in my loudest voice. I felt with my responsibilities as Quarter Master I stood a good chance of going on the short list for Leading Hand.

When I answered my girlfriend's letters I had given some hint that maybe I might have something special to tell her within the next few months.

One clear, sunny morning when the sea was like a sheet of glass, the destroyer flotilla of the Mediterranean fleet was out at sea under the command of the four-ringed Captain in his destroyer. He did not often have the opportunity of putting into practice the full tactics of training that is required by the individual ship's

Captain to train as a flotilla, manoeuvring together, as they were rarely free to do this. The Captain was standing on his pedestal in the forefront of the bridge, the Navigator at his right hand, and the Lieutenants were at their controlling position, along with the Yeoman of Signals who had his telescope to his eye, trained on our destroyer leader. The Yeoman's small signalling staff worked from the flag deck just below and on either side of the bridge, read morse or dashed around clipping on to halyards the answering pennants and flags, hoisting them to the yardarm where they fluttered against the strong breeze caused by the speed of our destroyer as it sped through the water. The propellors churned up the sea to leave white-flecked patches of cascading foam. The ships criss-crossed each other's wake as they swept abreast, then formed in line, racing full out, stems slicing the water as they slewed round, keeping their station in perfect rhythm. It was seamanship at perfection with all the traditional service skill and knowledge handed down from centuries. Modern machinery, gunnery, signalling, speed and range made split-second timing all-important.

The seamen hung on as the ship heeled over and the salt spray swept over the upper deck. The Chef cursed, his dishes slipping, lids flying off to give a loud tinkling cymbal, an accompaniment to the Chef's long outburst of unrepeatable words. A word of sympathy to the Chef at that time only led to another outburst. A blind eye a deaf ear and with time allowed for simmering down, were practised by all to our Chef in this time of stress.

In the wheelhouse an atmosphere of keen excitement was prevailing. The smooth sea gave no trouble. The ship answered to the slightest turn of the helm; even putting the wheel 'hard over', letting her slew round and then heel over on the opposite beam as I let the wheel spin back only added to the pleasant sensation. I had felt not the slightest concern these last six months or more when before I had bashfully and consciously kept my eyes glued to the compass. Now I could turn the wheel with one hand, talk at the same time, ring on the revolutions and turn to see whoever was present.

'Right, now.' The telegraphman dashed outside the wheelhouse watching the destroyers coming out of their turn together.

'Come back in. Stand by your telegraphs. We shall be altering course to port and changing speed,' I call, and lean out towards him. At that moment down from the voice pipe came the order 'Starboard 15'.

Was the order Starboard as leaning out I said Port?

I put on Port wheel. A few moments later a loud order: 'Starboard 30. Stop. Starboard. Stop both.'

The orders came down clear-cut and sharp as if some kind of pandemonium was taking place. A pause. then: 'Course 0.30. Both engines half speed ahead. Who is on the wheel?'

'The Q.M.' I replied, not being too communicative with my name.

'Send the bosun's mate for the coxswain to take over,' was the crisp order. The coxswain duly arrived to relieve me and I had to go before the Officer of the Watch to be asked why I had put the wrong wheel on. I replied saying that I must have heard wrongly and was told to go and wait in the coxswain's office.

I came down from the bridge crestfallen, and kicking myself for slipping up so easily. There would be a big enquiry I thought, waiting anxiously in the coxswain's office, and tried to think out what line to take. It was best I thought, to say very little, not to say anything yet about the telegraphman going out of the door. It would be part of my defence.

'Take it easy and calm yourself,' were the friendly words of the Coxswain when he entered the office. 'Tell me what happened now?'

I hadn't felt at all upset during the incident and had carried on steering some time after, before the coxswain relieved me. It was better perhaps to let him think I was upset and confused whilst resting in his office.

'I must have almost c-c-c-caused a c-c-c-ollision', I stammered, 'I c-c-couldn't have heard cor-correctly, and-and put the wrong wheel on,' I falteringly replied.

'Take it easy, son,' he said again. 'When you feel better join up with your part of the ship. I have to detail someone to take over your watch.'

I rejoined my old mess and asked what happened and if I had almost caused a collision. I was told nothing much had happened but one destroyer did come a bit close. Two days later I was asked by Stripes if the bottom had fallen out of my world.

'These last two days you have been silent and downcast. Do you think you are the first Q.M. to put the wrong wheel on?'

'I-I-I suppose', I stammered, 'there will be a big enquiry, and how can I but shoulder the blame?'

'If you ask me', he replied, 'you won't hear any more about it.'

'Why?' I asked.

'When doing manoeuvring, it is the Navigating Officer's duty to see that the coxswain is closed up at the wheel. Forget it. You won't hear another thing.'

I didn't hear any more, but when I turned up for instruction for higher promotion I saw the officer look my way and confer with the instructor. My name was called and I walked over as requested. I was told quietly and politely that it would be better for me to wait awhile. I saluted and turned away, disheartened, dispirited and disillusioned. I was prepared to take my punishment for my lapse, but I felt that a deeper wound had been inflicted that time could never heal.

During the latter months of our commission when we developed the feeling of being Old Salts, fully trained and acclimatised to the Mediterranean station, many rumours circulated round the ship that our flotilla would be relieved before the end of the month and that we would be sure to be home in time for Christmas.

We went to Malta, made acquaintance with the 'Red Garter', 'Blue Dolphin' and 'Silver Slipper', etc. Our dancing girlfriends threw their arms round our necks in a warm welcome back, but we were now more composed and shrugged our shoulders with a laugh, and gave them to understand that we still enjoyed their company but not with the first wild taste of careless enthusiasm. The honeymoon was over and we had become mature, overseas sailors. Thoughts of home and the thought of meeting loved ones again had a steadying effect. I felt I must begin to save. When Jolly Jack entered his local, his wide bell bottoms admired and a sweet young girl touching his collar for luck, he must be free, easy and generous.

Our statesmen were still making bad decisions and running our country wrongly. The mistakes made by these great men caused squabbles and upheavals and there was little consideration for the destroyers holding down or keeping check on the poor but temperamental states. There were outbursts and fresh eruptions from other countries, and Britain, if not directly involved, was so immersed in the political web of international affairs that we had to have a watching brief even in other foreign countries. When Spain erupted in Civil War, the Sea Lords, sitting in the polished conference seats of our Admiralty, suggested to the Foreign Minister, who put it to the Cabinet meeting, presided over by the

Prime Minister, that H.M.S. *Ardent*, a destroyer on the Mediterranean station, be given orders to patrol the Spanish coast from Gibraltar to Marseilles.

We were sailing at a steady twelve knots with our expected time of arrival in harbour piped round the messdeck. To everyone's surprise the ship suddenly did a 180° turn. The Chef uttered a yell as everything in the galley slipped. On the messdeck those off watch suddenly found themselves slipping, having to catch hold and hang on while the messdeck utensils with a clatter landed over the opposite side.

The trained sailors on the ship cursed the Q.M. on the wheel, and as the speed and vibrations increased the cups and plates clattered in the mess shelf. Retrieving the mess gear and making secure with the ship still maintaining top speed, the experienced sailor knew that something alarming had happened and that something of urgency was about to take place in the diplomatic world. We rushed on throughout the day and night to an unknown destination, with Chef declaring we had passed Gibraltar and would soon be seeing the lights of Southend Pier.

When we did slacken down our Captain let us into the secrets of the Admiralty. Our orders were to patrol along the Mediterranean Spanish coast from Gibraltar to Marseilles, picking up refugees from the Spanish ports. Extra lookouts would be closed up day and night as the Spaniards were at war and we might be mistaken for 'the enemy'. This must not happen and cause an international incident, so the White Ensign would be flown from fore, main and mizzen, with the ship's ports and deadlights down.

We felt this put us in a state of semi-preparation for action. Although we were ready to run our statesmen down there was a feeling of excitement to think we were in a war zone. It mounted when the Gunner's Mate went round in his official capacity, with a list of the gunnery branch who held the official rating of Seaman, Gunner, and detailed them off in pairs.

'You will draw and wear tin helmets, issued from the stores, and man the mounted twin Lewis guns day and night,' were his instructions. Circulated round and hung up on notice boards were silhouettes of Spanish and German warships. This was unexpected, and fresh news to us on the lower deck to learn that German warships were at sea, and even in the vicinity. This led to a lot of speculation, opinions were voiced, listened to and digested, for this was a serious occasion.

Our normal sea routine was having to be adjusted so that extra hands would be made available as look-outs and gun crew. We went into Gibraltar to refuel and to stock up with stores, and for the next few months we were to be kept running from port to port, doing much more sea time than any of us realised. As we arrived at and dropped anchor at these Spanish ports, boat loads of refugees would be ferried out to us. We were always anchored a considerable distance from shore, sometimes with a big swell running, or the sea was extra choppy, making it difficult and needing careful handling to get the boats alongside our small gangway. We would be on deck giving all the assistance possible with fenders, lines and a helping hand. The small makeshift ferrying craft were lifted high and low, battling against the sea, swept against the ship's side, then out again. These refugees had to pay with all their money or jewels to get whatever transport they could, probably finding someone quite inexperienced to ferry them out.

It was pitiful and heartbreaking for us, knowing the sea, to think of the rough passage they would still have to face on a slender destroyer. Those whom we assisted from these leaking boats ranged from tender babies, boys and girls to old and feeble men and women. The babies were wrapped in shawls which their anxious mothers clutched tightly, but tenderly as only a mother can. The angry waves would never separate them whilst life still breathed.

Lines would be thrown out and finally made fast in the boat bows and stern, whilst hands on board tended to and steadied the boat loads of refugees against the rolling ship and lifting gangways. Two Able Seamen stood on the narrow gangway steps with one strong arm gripping the safety line and the other one outstretched to grasp and hold their rocking boats, whilst other sailors gave a helping hand, lifting and passing them from these boats into the safety of the many willing helping hands aboard, who with food and drink and words of comfort tried to calm their anxious fears. Poor, trembling unfortunate people who had to suffer because of warring men, with greed and fixed ideals in their unyielding minds which could only lead to bloody war, death, destruction and misery.

Now that they were on board these poor people began to think that their salvation had come, but more hardship was still to follow. The Officers' bunks were filled by those in need, and the remainder were cared for by the ship's company for'ard. They were made comfortable with bedding and cushions in every empty space and corner that was available. It was uphook and away at top speed to the next port for more, causing our destroyer to pitch and toss as she raced onwards, for it was better to get the voyage over in the shortest time. Often the sailors were sick themselves whilst attending and running round with buckets for their sick and unwell passengers.

Some lay prostrate, as if dead, when we passed through the Gulf of Lions which is well known for its treacherous sea, always encountered before arriving at the disembarkation port of Marseilles. A look of relief would come to their anguished white faces as we watched them step on to solid land, and we were thankful ourselves that their rough passage was at last over.

We remained overnight in port and perhaps had a few hours leave, then we were off on the return route, Barcelona, Valencia, Alicante, Cartagena, Malaga and finally Gibraltar. Other destroyers also took part, with others engaged on the Atlantic coast.

Once when we were way out at sea, two dark objects appeared over the horizon, rapidly approaching. Their course would take them right across our bows. The upper deck soon filled with hands as word spread around that two large cruisers of the Spanish Navy were in sight. Eagerly watching with the feeling that these two vessels were stripped and ready for real action, they appeared black and sinister in the fading evening light. Onwards they came, getting even larger. Who would give way, we asked ourselves? Then we saw a big turret swing round, the long gun barrel pointed towards us. Being a Gunner myself I knew that men looking through binoculars had their sights trained on us. Already they had our range. I felt a nervous excited tension. I looked up to the bridge. The figure of the Captain stood undaunted, looking straight towards the oncoming ship. The white ensign flew from the foremast above. Would our flag see us through? Nearer and nearer we approached, with no slacking of speed on our part. More clearly we could make out the outlines of figures on their bridge. We looked again at the closing distance; there would be just about enough distance between us without either giving way. Then thankfully the gun turret trained fore and aft, the two cruisers swept past ahead bent on their voyage of destruction. I watched them fade out of sight with a realisation that life in the Navy would not always be carefree and peaceful. We too could be sailing on a mission of war.

There was a night when a voice, as if in distress, was heard above the noise of the rough sea that was running at the time. We were lying at anchor off Cartagena. It was a dark, blustering night with the stars hidden from view by the low threatening clouds. The alert lookout stared into the inky darkness as again the shout of distress faintly reached us. Quickly the light of an Aldis lamp was beamed on the water. Struggling against the waves we saw a figure, not twenty yards away. Those last twenty yards for this courageous swimmer were a life-and-death struggle. A lifebuoy with a line attached was thrown. We watched the beam of light played on the swimmer and the lifebuoy. The waves swept them apart, then brought them closer. The battling swimmer with life and strength ebbing away each second tried to reach the buoy as it was lifted high, then pulled down by the rough sea. Anxious hearts beat fast with all their concentrated praying going out to that lone swimmer. 'Would he make it?' The waves lifted him up to show his aching arms flailing the water, his tired legs kicking out. The small number of watchers now leaned forward as if to give him that extra energy. The rest of the ship's company were unaware of this life-and-death drama happening so near to them. He must reach it this time we prayed, as the swimmer and the buoy reached the same rising wave together and then the shoulders and head of our unknown hero disappeared, just as safety seemed so near. A groan of despair went up from all, with muttered oaths heard in choking words. Then, as if our unspoken prayers had been answered a hand came out of the dark water, a head bobbed up and an arm groped round the floating buoy. A cheer of delight came from those throats that a second before had closed with grief.

The young, dark, handsome Spaniard who had reached the safety of our ship stayed on board until reaching Barcelona, where he wished to be landed to rejoin his comrades. The story behind his heroic swim was that he was spying until challenged. He and his comrade had fought their way out of a building where they had taken cover, knowing it was only a matter of time before they would be captured and shot. They had seen our ship, although they did not know its nationality. They had decided that it was the only hope left to them and they had made a break as soon as darkness fell, jumping into the sea and trusting their life in God's hands. They both knew that with such a sea running their human effort alone would not suffice.

Almost two years had passed since leaving the shores of old England. It seemed a long, long time to be away, cooped up in our shell of iron, waiting for those last remaining months to come and go, but the more you yearn, the longer the time. Perhaps life for some becomes unbearable and tempers get out of hand. One afternoon, when hoisting barrels aboard from a derrick, rigged up on the port side of the fo'c's'le and attended by a dozen hands, with an officer standing by, I felt full of life, jumping and acting in good spirits, unaware I was upsetting anyone as I hauled on the rope and dropped it quickly to the order of 'Lay to'. I didn't know my high spirits were so annoying to one who was feeling down in the dumps, for as the order came to pick up the rope I enthusiastically jumped to obey and received a hefty punch on the jaw, aimed at me so deliberately that I was amazed, and stood as if rooted to the deck. It wasn't the sting of the blow, but the surprise. I didn't do anything, just stood there until I was moved by others hauling away. It was so quickly delivered that the Petty Officer didn't see it. By the time I recovered from the shock I was hauling away under the officer's eye. Anyway, I had time to think. What was the sense in hitting back and both getting into trouble?

'What did you do that for?' I asked the rating.

'I just felt like it,' was the reply.

I didn't do anything about it, but it was mentioned at the tea-table and I noticed one or two looked at me under raised eyebrows when I said I didn't retaliate, and I began to notice a falling off in their friendliness. Was I mistaken or were they thinking I was a coward and lacked courage to stand up for myself?

This worried me far more than the blow did. I didn't want to be looked down on in that way, to be shunned and despised by my mess chums for lack of courage. Two days now had passed by and I fancied the whole messdeck were eyeing me contemptuously. Didn't the Bible say 'An eye for an eye and a tooth for a tooth'? He had punched me without any warning. I would return the blow and see what he did. This thought hammered in my brain, not that I wanted revenge on the rating, but to retain my shipmates' prestige, which I fancied I had lost. I watched him depart from the messdeck. I went down the passageway to catch him on his return. I saw him coming and walked pass him. Wham! I let go my fist, catching him well and truly on the chin.

'A blow for a blow,' I remarked, as he uttered 'Oh', putting his hand to his mouth.

'On the fo'c's'le, on the fo'c's'le,' he yelled out and rushed on the messdeck to 'off clothes'. 'A fight, a fight', he roared out. It was taken up so quickly that men ran past me as I walked into my mess. The stokers were coming up from the mess hatch below, all making for the fo'c's'le, whilst I, a bit bewildered, stood in my mess, beginning to realise that I was the cause of all the excited rushing of the for'ard ship's company.

'What happened?' gasped a mess chum.

'I gave him a blow for a blow – he hit me the other day and he yelled on the fo'c's'le "off jumper".'

'I'll be your second.' I was helped off with my jumper and singlet so that I stood in my trunks and slippers.

'Come on, you are ready,' he said with such excitement in his voice that you would have thought I had won a prize instead of being an opponent in a bare-knuckle fight. Stripped down to the waist, being urged to go as all was ready, I walked down the passageway up the ladder to the starboard screen that led to the fo'c's'le. Men made way to let me pass, eyeing me and weighing up my chances compared with my opponent I supposed. What made it worse was that we were tied up beside the *Anthony*, another of our flotilla. I saw men lined up, shoulder to shoulder on the guard rail, climbing and swarming over the bridge, with the Chef and the P.O. leaning over the 'B' gun deck. Being the centre of attraction for my rash deed, I didn't feel brave or a coward, nor did I want to fight, but I realised that now I must, for all eyes were upon me. My opponent was standing in the only clear space on the fo'c's'le. He was also stripped, with fists clenched.

I stepped to meet him and heard a ruffle of excitement from the onlookers. We ducked and circled as our fists flew. I kept sticking my left into my opponent's face, then covered up or ducked to keep away from his blows. I felt I was doing well. I could see his lips puffing up as he gasped again and again. My quick left went into his swelling lips. He swung as I covered up. We were both tiring and then he caught me. It came in underneath my arm, right full in the eye, and I saw stars. I lost my head as I staggered back. Now I was deaf to the crowd and only saw my enemy as he advanced. I must act swiftly, something told me. I leapt in, stung into action by the blow. My left was in his face again and again. I mustn't let him get another in. I felt him going back and now he was leaning on the guard rail. He was so tired I used my right. He covered his face with his hands and leaned on the guard rail as I delivered another right.

'That's enough,' my mess chum said, catching hold of me. Heard the crowd's feverish applause and realised it was over. I had won, but not without the scars of battle. My eye was aching and bruised and puffing. I felt it as I put my hand up to wipe away the blurred vision. My enemy came towards me to shake hands.

'Did you offer your hand after you hit me two days ago?' I asked him. 'Go away. Be off', I said as if I had more hatred now that it was ended than when I started. The incident had brought the whole ship's company for'ard, to be the leading topic for the next few days.

Returning to the mess with all my messmates looking at my now closed eye they offered advice.

'What you want to cover over that is a large beef steak. That would stop the bruising.'

'You're going to have a beauty in the morning.'

'You've got a good left. You shouldn't have ducked so much.'

'It was a good fight. Better than going to a boxing match. See the puffed-up lips of the other chap. He won't be capable of blowing his trumpet for Sunday service. Jimmy will be annoyed. He thinks a lot of his three-piece band playing at church service on Sunday morning.'

I didn't feel all that great, even if I had won. No doubt I had won back the respect in the eyes of my shipmates, but with this big shiner that stared at me from the mirror the following morning, after having to stand to attention for inspection I didn't feel so good. There was no way of hiding the black eye that seemed so large as my Divisional Officer stared and then passed on. I carried my scar a week before it faded away.

'Are you still going strong with your girlfriend?' asked Jock, my Scottish pal.

'Yes,' I replied, rather boastfully. 'Her letters to me have never failed.'

'You're lucky,' he replied, 'my girl friend's promise only lasted three weeks, and then she wrote that she was sorry to have to tell me, but it was better for me to know now, that she was going out with another boy. If I cared to, she would carry on writing as a pen-pal. I never answered, so I shall be a complete bachelor on my return.' He spoke regretfully, as if I had something precious which he had lost.

I replied, 'I think she did the right thing. She did have the decency to inform you right away. You never know, when you get back, you may pick up where you left off. Is that her?' I took a closer look at the photo of a smiling girl that had top pride of place in the centre of the lid of his ditty box.

'Yes, that's her.' He took it tenderly from its position for me to view it closely.

'Oh, she is a sweet bonny wee lass,' I say to him, as I hold it and look at the picture of his Scottish lassie.

'Why don't you write just out of the blue and tell her you will be home shortly?'

'Jock, it is time to think about getting some souvenirs to take home. I have a good mind to get yards of silk or satin which look so lovely spread out in the shop windows. I must take everyone something. Besides, if I buy some lovely white silk as a present for my sweetheart, who knows what use she may make of it?' I gave Jock a dig in the ribs. 'Pure silk, like the Arabian princesses wear, brought from the bazaars of the mysterious East to make into a wedding dress by skilful hands, cut to shape, with a train so long it will take six bridesmaids to hold.'

'Don't you plan too far ahead. Remember our commission isn't over yet' was the advice my canny Scots friend gave me as he saw my eyes light up with enthusiasm at my thoughts. Tomorrow we would be back in Gibraltar where all the merchant ships called in to unload their Eastern treasures, where the merchants buy them not only for the tourist trade but also for the rich Eastern Arab sheiks, whose small kingdoms are far over the Sahara desert. These chiefs send their buyers to this port once or twice a year.

That night as I lay asleep in my hammock, rocked by the gentle swinging of the ship, I saw the grey towers of an English church, heard the bells ringing out, the sun shining on my sweetheart and pictured her beautiful English loveliness in a dress of the finest Eastern silk that glittered and rustled. She carried a bunch of pink roses and was attended by six little girls all in pink holding that long silken train.

No wonder I woke up that morning feeling contented and reserved, keeping quiet, for my head was full of thoughts of what I was going to do. This was a job for me alone. All on my own I would go ashore to look round and buy those treasures to take home. My sweetheart would never dream of what I had in mind, I thought to myself as I went about my duties. There would be her letter in a blue envelope, with one from home in a white envelope. I would always read this white one first, leaving the precious one last. After the ship had been secured and our refugees had disembarked, I walked into my mess. There were my two letters, amongst several others, lying on the table. Some of the lads were already seated, deep in the contents of their news. I picked my two up, took a seat and opened the one from home first. I smiled as I read of all the little incidents that had happened. It would not be much longer before I would be stepping over the threshold, I thought, as I laid it down to finger the other. The other letter I always read again and again. I carefully opened it, taking out the folded sheet and settling back with my head full of the nice plans I had in mind. What was this? – I started again – one thin sheet, beginning with 'Reg'. Always before it had begun 'My dearest darling'. I started again from the beginning; 'Reg, I know you will think the worst of me, but is it worth while writing to each other, all these long empty months wasting our time? I know it is bad of me, but please try to understand won't you? Our nice holiday together I remember, all the letters you have written, I have them all, tied together by blue ribbon, but please understand the hopelessness of our lives with you always having to be away. It would be better for us to finish and say "Goodbye". I don't know what I shall do in the future. I have no plans. Goodbye my darling for the last time. I wish you a happy future. Goodbye. Isobel. X'

I stared at it and reread it over and over again. Banished from my mind were those enterprising plans of a few moments before. Deep thoughts filled my mind as I sat still and silent, unaware of life around me, for I heard and saw nothing in the few minutes it took to sink in. For me to realise that at last it had happened to me. All the warnings and the sneering remarks of others which I had so patiently endured. My faithful trust in her; did it lay shattered and lost for ever by this sheet of thin paper I held in my hand? I collected myself, brought back to reality by someone pushing past me. I picked my letters up and walked out, wanting somewhere quiet where I could be alone to think it over. I went up the ladder on to the fo'c's'le deck, up the ladder to the wheelhouse, right on up to the upper platform of the bridge itself where I was alone and I could hear anyone approaching from beneath. The rock of Gibraltar looked down on me. Spain stretched from the foot to distant hills and mountains, whilst across the straits the gloomy outline of the Atlas mountains of Africa lay bare and rugged. I looked out into the distance, deep in thought. I unfolded the letter again and reread it, digesting every sentence before putting it away. I stood, chin cupped in my hands, until I became aware of the blue sea, with small ruffled waves sweeping onwards, never still. I must carry on the same, and keep it to myself for the time being. All was not lost; what sort of chap would I be to give up, to admit defeat? I would go ashore just the same, but not to drown my sorrows in drink. I was made of sterner stuff than that.

I went ashore, walked past the bars with music floating through the open doors, into the shops and bazaars, eyeing all the nice things, until attracted by a beautiful teaset. I entered and bargained with the proprietor. Perhaps it was my temperament or determination, or my recent upsetting news, for no agreement could we reach. I was adamant on my ridiculous cut price which I insisted was all I would pay. The price asked was 36s, I offered 14s. I walked out although the price had been knocked down to 22s, back to the ship and supper which was always at 19.30. Not until after supper did I take pen and paper to write my reply. I would write no bitter, hasty words, would not appeal to her, but I would write a straight and honest reply.

'My true love,

You will always be that to me, no matter what you write in your letter. I have read and reread it, and thought and thought. Can you honestly put away for ever that deep understanding and feeling which we had between us? It had developed into a binding love which I held sacred. Our slender chain of weekly letters has stretched across the ocean, linking us together. How I looked forward to them. After the sea trips of these long months, when we returned to harbour, I searched for and found that blue envelope. Dearest, have I to return from my next trip to find that binding link of our love missing? No letter for me to pick up? Not to be able to sit down and become unconscious of the others around me whilst I read the loving words you have written to me. No my dearest sweetheart, I cannot take your last letter seriously. I have not told a soul and I am going to carry on writing now as if it never came.'

I finished my letter with the news and endearing words as

usual. I posted it and attempted to carry on in my normal way. I went ashore again the next day to carry on the bargaining for the teaset. I had my mind still firmly set on 14s. The proprietor called me a 'Scotch Jew' when he came down to 18s. and I still refused and walked out. I had not gone far when a young boy tugged at my sleeve and said 'Mister 16s'. I relented and returned to pick up my bargain. I have it to this day. It is given pride of place in a glass cabinet.

Our next trip was along the Spanish coast. I felt my fate rested on my last letter. I would not hear for ten days to a fortnight. I tried to be myself, but for long spells my thoughts went flying across the sea, absorbing my mind as I re-lived our times together, our walks when first our love was awakened. I remembered our linking hands when the shyness wore off, the night I walked back to barracks after our first kiss and tender embrace, when I had almost danced down the road wanting to throw my hat up to the moon. The effort I made to restrain myself. Our nights off from duty – how eagerly I looked forward to sharing a few hours together. The happiness when we found ourselves on the train with a whole fortnight together before us. Going to my village home. How proud of her I was. The train rocked and roared down the track and her eyes shone as she told me how it brought back childhood memories of her home in Canada with the rail track running by. Her shyness as I introduced her to my parents, and I had held her hand to give her assurance, my place by her side to give her confidence and protection. All these memories came flooding back to me as I drifted into a dream of our romance. Was it to be broken, gone for ever with just a memory left to me? 'It is not going to be,' I would think to myself, and put extra effort into whatever I was doing. Reverses only stimulated my determination. Thank goodness the trip was filled with incidents. If they did not fully succeed in taking my mind off my dilemma, at least it stopped my messmates from observing my predicament.

We met up with some German warships and challenged their crews to a game of water polo. It gave me a chance to observe these Germans who had sailed their Grand Fleet against the might and seamanship of the British Fleet less than twenty years before. The German ships, anchored not far away, were modern and streamlined and kept to a high standard of cleanliness. They were more of a showpiece than our weathered destroyer, but then we had been working hard, engaged on our mission of rescuing the Spanish refugees.

Their sailors were blond, muscular youths, the pick of the German fatherland. They were well behaved and highly trained even in the water, ready to commence the game. They looked the more powerful side and their orderly formation looked set to overwhelm our team. However, our slender-looking lads dived and weaved and twisted, giving touches of individuality which threw out of line the skilled formation of the German teamwork. We won by an odd goal.

I wondered if I had been watching the true hereditary characteristics of both races showing so clearly in the game, but which could also be a true strategic reflection of a future war.

One morning I witnessed the first firing of guns in war, when the harbour echoed with loud explosions to our surprise and curiosity. We were working on the upper deck and heard a large number of guns popping off in remote positions on hills and by the roadside, as well as close by. We wondered if they had mistaken us for an enemy, but then someone spotted a plane, so high it was scarcely visible, which dropped some bombs on the town. As the gunfire died away this plane made off and we observed that the shell bursts were very inaccurate.

On our return from Marseilles we heard that a Spanish plane had mistaken a German warship for one of their own rebel cruisers and flying low had dropped a bomb on its quarter deck, killing 18 German sailors. The following morning, at daybreak, the German warship lined up off Almeira and bombarded it for two hours, killing or injuring 200 inhabitants in reprisal.

I lost some respect I had gathered for the German race as I could not see our Heads of State giving such an order, knowing that women and children would be killed, in revenge for sailors' lives lost in a mistaken-identity action. This was a risk which had to be taken by a country when patrolling another's coastal waters, knowing they were interfering in a civil war.

One forenoon watch we were sailing along in the ordinary way when, looking ahead, it dawned on me that the clouds above seemed to be getting low. There were weird streaks of blackness against the greyish tinted clouds which were billowing out like a balloon being blown up. I looked down at the sea. It lay calm and blue with a seagull gliding above, its curved wings extended to catch the lifting current of air.

Although there was this calm, I began to feel that there was something unnatural about it. For some reason we spoke more quietly. I was on the wheel, having been restored to my old job of Quarter Master. 'Look ahead,' spoke the Bosun's Mate, 'the sky has turned black, with greyish threatening streaks sweeping downwards, and it is building up all across the skyline.'

'Bosun's Mate,' orders the Officer of the Watch from down the voice pipe, 'inform the Duty Petty Officer he is required on the bridge immediately.' 'Aye, aye Sir,' I repeat, passing on his orders, then looking round myself to see if everything is shipshape as if preparing to meet some unknown emergency. 'Close both doors,' I order my telegraphman, 'screw up all ports. To me it looks as it we won't know whether we are airborne or sailing when we run into that wall of greyness.'

I look over the instruments. A straight course is being kept. Both engines are half speed ahead. I turn the wheel slightly. No worry, everything is normal and she is handling all right. Already angry wisps of fraying cloud hang precariously in mid-air, streaks of pouring rain sweep round in a whirl of wind. The surface of the sea becomes ruffled and black as the threatening clouds hide the sun from view.

'Quarter Master.'

'Aye, aye Sir,' I answer.

'Secure the wheelhouse, switch on your illuminating lights. Be prepared for a buffeting.' I look round for a second time and report back that all is secure in the wheelhouse.

The Bosun's Mate returns, informing us that the duty watch are turned to securing boats, cables, watertight doors, hatches. Everything is being screwed, lashed or battened down. Even Chef is given a line to lash himself in his galley. With a rising, tearing roar the wind strikes against the ship. I feel her shake, then the whipped-up sea makes her pitch. I grip the wheel more firmly, trying to keep her steady on course, but tearing gusts are hurled against her upper work, whilst the sea is whipped to a fury. Wild, angry waves lift and carry our small destroyer, and she bucks and rears, being caught in the grip of an angry turmoil of surging water which she gallantly struggles against.

'Look', shouts the Bosun's Mate, 'those curling streaks whirling around. It's a waterspout right in our path.'

'Starboard 30', comes the order.

'30 of starboard,' I repeat again when the wheel is over.

'Steady.' I read off the degrees. 'Keep her steady' comes the order from the bridge. I wrestle with the wheel, trying to keep her on course, for the wind is tearing and roaring, the sea is a heaving mass of whipped-up waves with white foaming tops curling angrily as they roar their defiance.

I see the fo'c's'le rise before me, up and up, then pitch down. I wait for the thump and shudder which I know will come as she strikes, then right herself to rise again to meet the next wave. It seems as if the ship is battling against some unseen force with an enemy that has tormented the sea to its very depth.

'Look,' exclaims the Bosun's Mate, alarmed, 'there's another waterspout to starboard.'

'By Jesus, the fury of the Gods be upon us,' cries my Irish telegraphman.

Sheets of rain are beating against the bridge, hiding the fo'c's'le. No longer can anyone see from the wheelhouse.

'Port 30' I faintly hear down the voice pipe although I know the officer is shouting at the top of his voice. I repeat, putting the wheel over and wondering how she is going to act as the degrees spin round. The ship rises up, then plunges, heeling over as she turns. The Bosun's Mate goes sliding across the wheelhouse, the sliding doors shake and clatter. For a moment she seems to hang suspended in mid-air as if about to plunge to the bottom. I feel sure the stern is lifted clear of the water with the propellor churning the air, then she dips and I hear the waves above the wind and rain crash down on the fo'c's'le. She shakes and shudders under the impact as we all hang on to keep upright. With relief, I feel her rise, going forward.

'Steady' is the order, and I yell out the degrees, fighting with the wheel to get her on course. The wind and rain lashes and moans round the wheelhouse whilst the ship rises and smacks down again, shuddering as if shaking before meeting the next onslaught.

'Keep her steady' is the order, yelled down.

'Am doing my best Sir', I shout back, as I turn the wheel to port or starboard, trying to check her swing, as she flounders in the unrelenting force of the storm. Then a break comes in the roaring wind, the ship straightens herself, becomes more balanced, her steering steadier. The rain stops, the wind dies away, visibility is again clear. The fury of the sea is over, only a choppy surface remains with streaks of sunlight streaming down from the broken clouds above, and it is all over.

'I'll bet our crockery has gone west,' I remark. 'Do you know what?' I ask, 'We must have been dodging between waterspouts. It couldn't upset our old tub' I cry out exuberantly, giving the wheel a resounding smack, declaring 'She is British through and through'.

At last the day arrived when Gibraltar loomed into view. The trip seemed so long as each day I thought and wondered about a letter and whether one would be waiting. The Civil War, the German ships and the waterspouts at sea were real incidents but insignificant to the thoughts that occupied my mind. I felt that my whole future rested on whether a blue envelope would be waiting for me. I tried to get a grip on myself, to remain calm and steady, and now at last we were gliding in to port. Soon, now, I would know the answer to my anxious thoughts. I patiently bided my time and kept clear of the mess until the mail had been sorted out. At last I walked on the messdeck. Dinner had been served up, with men taking their places, sitting down to eat with the others, reading or pocketing their mail. I glanced down at the remaining envelopes scattered on the end of the table. How casual I tried to be, but how eager was the glance hidden beneath the casualness. Fingering the letters to one side my anxious gaze swept from one to the other, and there it was. I steadied myself, taking a second to pick it up as my heart hammered with excitement, and as I turned it over I saw a large X on the folded flap. I put it in my jumper next to my thumping heart. I would not read it before my chattering messmates. I wanted to be alone. The big X and clear handwriting raised my expectations. I had waited the last 14 days at a low ebb, now my hopes had risen. The stimulation of a few minutes would make amends. I took my regular place at the dinner table, giving my mess chum a nudge to make room for the one who had kept the ship on course, sailing between waterspouts.

'Ho, ho,' he laughed, as if my boastfulness was nothing. 'What you mistook for waterspouts was the Chef emptying a can of water over the side.' He gave a wink and smiled broadly at his messmates.

'Some can of water', I retorted, 'when it took the Stoker a whole day to bale out the water that poured down the funnels. That beat your typhoons in the China seas.'

'What you encountered only lasted five minutes,' he answered back. 'A typhoon would have raged with a terrific wind for hours.

There wouldn't have been a mess table standing or any crockery left.'

Although I had been feeling exuberant since picking up the letter, supposing I was wrong? It could still contain bad news. I put my hand in my breast pocket, feeling in it lovingly as I stepped out of the mess unobserved, I hoped. I made my way up to the bridge, taking it from my pocket, and settled down on the Captain's stool – a worthy place to receive important news. I tore the envelope open and read:

'My dearest love' – I smiled with pleasure at those three words. It continued:

'I am sorry, dear, that I wrote that letter to you. Please understand, my darling, that I was feeling so miserable and heartbroken, and it seemed so long since your farewell kiss. I get so very lonely just longing and writing, wondering if you ever will come back. I was so very pleased to get your letter saying you would take no notice of mine. I love you now and always will, but please darling hurry home to me.'

I put my letter down and stared at the painted bridge through a cloud of mist that blurred my sight. How could I help my weather-tanned cheeks becoming moistened by watery eyes? I brushed them away with the back of my hand. Tears of happiness choked me as I felt my deep feeling of love so overwhelming me travelling across the miles to my love across the sea. I stood up without reading more. I must go down to the mess and write an answer to assure my sweetheart that I loved her more now than ever before.

CHAPTER NINETEEN

It was a happy ship's company who lined the fo'c's'le to have their photograph taken. All the officers and men who served on H.M.S. *Ardent* were there. An official notice had been posted up saying we were expected to arrive home early in April after two years of overseas service. We went about our work smiling readily. Contentment prevailed amongst everyone. We could all now count the days and post the good news home. Inwardly it brightened one's whole disposition. Presents of all kinds were bought and stowed in lockers and secret places, as space was so limited. We spent many happy hours amongst ourselves gathered round the mess table or lying talking in our hammocks of what it was going to feel like lying in a four-poster bed. Stripey told us of his previous commissions after three years on the China station, arriving home and finding it impossible to sleep. He said he had had to get his wife to throw buckets of water against the window to make it sound like the waves splashing against the porthole before he could drop off. How a shipmate of his had arrived home to find his family had increased – an extra one for each year he had been away. He told us of a young seaman who had made an allotment to his sweetheart, sending money regularly. Passionate letters had passed between them. All was arranged for the banns to be called in the local church for the first Sunday after his arrival home. Stripes himself had helped him to pack the silk for his bride's wedding gown. This had been received and a letter thanking him had arrived, together with a photo showing the wedding dress being worn and a special request for some more for the bridesmaids. His messmates whipped round for a wedding present thinking that whilst most of their romances had broken down, here was one which touched the heartstrings, which had continued through three long years of foreign service. Now their happiness was to be fulfilled in the three months which it would take the ship to arrive home.

After this, no more letters arrived on the long voyage home. At each port of call there was nothing, but he was not unduly worried as this was understandable back in those days. Anything might have happened, an accident, or she might have been taken ill. By the time they reached home the shipmates could detect the effect worry was having on this steady and faithful sailor. They were so concerned that special arrangements were made amongst themselves that someone travelling his way should keep an eye on him and check that the mystery was happily cleared up.

When this young sailor eventually arrived at the house of his betrothed, there was only an empty house, bare windows for all to see through. On enquiring at the neighbours they learned that a wedding had taken place some two months back and the happy couple had left for London. That was as much as they knew. This was confirmed by the Vicar of the Parish when shown the photo of the sailor's sweetheart in the wedding dress made from the silk which he had so tenderly and lovingly packed. His travelling shipmate had tried to console him, offering to stay with him and inviting him to spend his leave with him, but he just wanted to be alone. When next they heard of him after their six weeks' leave had expired and he did not return with the others, they learned he was in a Sanatorium with a wasting disease that no doctors could diagnose. The one or two who went to visit him could bring no brightness back to his sad eyes, or a smile to that sorrowing face. He only lay and stared at the photo of his lost sweetheart in the wedding dress standing on the small table by his bedside. He died a few weeks later. The cause of death that was given on the Doctor's Certificate was 'Unknown tropical disease', but those who knew better knew it should have been 'Died of a broken heart'.

What a wonderful feeling sailing home is, each watch bringing you nearer to the shore! I leaned on the guard rail, standing gazing across the water to catch the first sight of the white chalk cliffs showing so clearly against the rolling countryside. Many ships are sailing to and fro. You feel excited, but your roving life helps you to contain that feeling. Your long months away and the stormy seas are all forgotten. It has all been worth while in the glow of enchantment that is sweeping all over your body.

You talk about your home town, the train to catch as you press your No. 1 best suit. I think of my love waiting. Will she look just the same? How shall I greet her after this long time?

The last night at sea, the ship sails steadily on. No longer after this will it be the Old Ship's Company any more. Each will go their own way to their homes scattered all over Britain. We may meet up in barracks after this ship's pay-off, where one small ship's company is soon absorbed amongst thousands of others that are constantly changing, joining and being drafted to all corners of our far-flung Empire. Great activity is carried on from this Chatham naval base which we call home. The paying-off pennant is trailing and fluttering in the wind as we steadily steam down the Medway. We can see the red brick barracks, then cautiously we are guided through the bull's nose, the gateway into the sheltered basins of the Royal Naval harbour, and are safely tied up by wires to the land bollards.

All our long-suffering months of Foreign Service are forgoten as we get ready, almost diving into our No. 1 suits. The first liberty boat is for those who are off watch. There is the ring of home in the sound of our tread as we hurriedly make for the dockyard gate. Our shipmates watching know that in the next day or two they will be doing the same.

Now we are through the dockyard gates. A few wives and relatives are there to greet their husbands and sons. I see one of my chums locked in the arms of his wife, who I think is weeping for joy as her head is buried in his sailor's collar. I hurry away quickly; we English always become embarrassed when confronted by an emotional upset, even if it is one of joyous homecoming.

I had written and told my sweetheart that if my ship arrived home by the date given I would be ashore by the first liberty boat and I would meet her at the usual meeting place, the same one of our first arranged date, the courtyard in front of Gillingham Railway Station, round about 5 o'clock. I had written 'You can forget the two years in between and meet and greet me as if I met you only a few nights ago. Pretend there has been no long interruption, for perhaps I am bashful and uncertain how I should greet my love after all this long time.'

I find I have to hang out time, strolling leisurely along, watching the faces of the many people as they walk by, seeing the local buses picking up and dropping down their fares. It is the old familiar England. Its shop windows, the names of the pubs, I remember them all as if it were only yesterday. At last the tall clock in the High Street shows five. I walk a little faster to the corner, cross the road, look up and there to behold is my sweetheart standing out from amongst all the passers by, looking so demure and bemused, with a slight smile of shy welcome on her adorable face. I too am becoming slightly embarrassed as I stand before her and smilingly say 'Home at last'. I tenderly kiss her, softly and gently, as we both suffer from Anglo-Saxon embarrassment and will not show our deep affection in public. I stand back to admire her stylish light blue coat with white lapels and little velvet design on them. A dimpled smile plays on her face. I see the warmth shining in those soft eyes that are looking at me in a shy and timid way. I take her hand as we walk along for her sweetness has really taken my breath away. I can hardly speak in this moment of stress.

'Did you miss me?' I manage to speak, but it is almost a whisper.

'No, you told me to forget those long two years, so I'm thinking it is only the other night that you were saying "Goodnight".'

I listen to the sound of her voice, the sweetness of its tone. I want to hear her talk. 'Tell me all about yourself.' I want to say as little as possible for I realise that I haven't heard an English girl's soft voice for two years, and it sounds so nice that I wanted to keep her talking to me in that happy girlish ringing voice, filling me with pleasure. As we walk along we turn a corner, a short street with no windows, and we are alone. I stop and draw her closer, looking into her eyes to see the warm true love shining there as our lips meet in a long, lingering, homecoming kiss, and then she is laying her head on my sailor collar and I lay my cheek on her soft hair as my sweetheart is clinging in my arms.

'Let's get moving, someone will be coming,' she says, and we move along slowly with our arms linked round each other with all the embarrassment and shyness falling away. There is so much to say, to think about and to plan. 'We must arrange our holiday together. Nothing is going to keep us apart after waiting two years, for I have my overseas leave to come and I just won't go home without you. It would be unbearable.'

'I know, dearest,' she replies. 'for haven't I been waiting and longing just like you? I want to come with you. I have been cooped up in my small world with two half days a week to break the monotony, when it has been the flicks, or an adventurous visit to Chatham with my sister, watching all the sailors and wondering if you would suddenly appear amongst them, but you never did. I just had to go on waiting and waiting. I thought the time would never pass. I got ever so lonely at times, just me in my small room. I would lie in bed watching the moon through the window panes and wonder if you were on watch, sailing across the sea and looking up at the moon, just like I was doing. I thought, that is the only thing we can both see together and I used to send my thoughts out to you via the moon, wishing it would shine down on your ship, showing you a silvery path to guide you back home to me. I'm being silly and sentimental don't you think?' she asked.

'No, you are not,' I replied, 'for I too many times have watched the moonbeams dancing on the waves as we sail along and they always turned my thoughts to you. Hasn't it fulfilled our wishes and brought us safely together. I am really home or am I?' I ask her. 'You are making me feel so happy I want time to stand still. Go on, tell me more of what you thought and did the long long time that I have been away. I can hardly believe that time has gone. Am I going to wake up and find I've been dreaming?'

She laughingly tells me, 'It's no dream.'

I reply, 'I've come back to stay and I don't think you are ever going to get rid of me any more.'

'I don't want to,' she answered. 'Where are we going?'

We had been walking along, so enraptured with each other that I found we had strolled down to the old seat under the tree by the river.

'Did you lead me down here?' I ask her, 'Or is it that silvery moon guiding us to our destination?'

We both sit down, this time without a space in between us as when we first met, but close together as lovers do. After I had kissed my sweetheart she asked me to tell her of all my travels, about the foreign countries and towns which she never expected to see. The hours passed so quickly, and it was late when we left our river seat to say 'Goodbye' at the gate of the house where she worked. Our hands lingered, then she dashed along the path and I was left alone to walk back to my ship, a happy bemused young man, my head singing, the world ringing, for I was head over heels in love.

Whenever possible I was ashore meeting my sweetheart. There were the presents I had brought her, the half-pint bottle of scent, duty free, that every sailor brings home, the flowered Eastern patterned kimono of the silkiest silk, beautifully draped, which swept down to the floor with eastern patterned flowers so brightly coloured that they looked more beautiful than the fresh garden ones. Satin tablecloths, sparkling clinging waves of softest fabric glistening with colours as they rippled into place when you swept your hand over them and, of course, the treasured teaset, so delicate and fragile, that we hardly dared touch for fear of breaking. All these treasured gifts did I bestow upon my precious love, who righteously declared that they would be taken care of in her bottom drawer, and last of all was the parcel of silk all folded up in creamy white wrapping and tied around with ribbon.

'This is specially for you,' I tell her, 'and you are not to open it till I have gone. One day, perhaps in the near future, you may have a very special reason to make use of it.' She took it tenderly from me, smiling as she gazed down at it, guessing, I knew, what it contained, but pretending she didn't.

'Now, I wonder what you have in mind bringing me a mysterious parcel that I mustn't open until you have left. Knowing you so well, there is something behind all this. What intentions do you have, some binding promise if I accept? Some Eastern spell will befall me if once I open it and don't fulfil the wish, some magic charm that may bind me to you for ever. Answer me, my sailor boy, or you can take it back. I am not going to be obedient and veiled to be kept from view. Those foreign ways you have learned from being in a strange land.' She looked at me suspiciously and frowned as if to make me believe that she didn't want to take anything from me that wasn't honestly given without any ties, and made as if she wouldn't accept.

I pretended to be angry. 'Give it to me back,' I cried, as if stung by her resentment.

'No you don't, you hasty tempered beast.' Now she was looking at me coyly, her soft eyes making me all gooseflesh as I dropped my assumed temper and seized her boldly and kissed her warmly, telling her she must mind her ways and not pretend to hurt me as one day we might both take it seriously. All she said in between my kisses was 'All right, I may not if you let me go, perhaps I like you best when you are stung into action.'

My head was now filled with romantic bliss, for all was arranged for my sweetheart to spend a fortnight, her summer holiday, with me. Together we would travel again on the trains taking us to my village home, on which, next to my love, my thoughts mainly dwelt. We had been a small destroyer and the men from the ship's company who had been closely packed were now waiting around in a spacious barracks, talking gaily with that bright eager look showing clearly on our features and in our bearing, waiting for clearance chits to set us free from the strict routine of Naval Barracks' life. At last we are swinging along, our small suitcases carried in our hands as we walk through those barrack gates into civvy street. We broke into a run and swiftly became scattered in all directions as we caught buses and taxis and hurried off in the pleasing excitement of our freedom.

At the station my sweetheart was waiting for me, dressed in as trim a costume as ever you wished to see, with an air of pride in her appearance that would make anyone respect her. I became aware somehow that a certain stage in our life had slipped by, for my teenage love had grown into a young lady during the time I had been away. And it was a different son who presented his sweetheart and himself to his mother when he was home from his foreign travels. 'You have changed and grown up so' was her remark after the greeting ended. I stood back for her to look me over, seeing a mother's pride glowing on her worn face at what she saw. 'Yes,' I replied, 'I have grown up into a responsible young man.' I was so pleased with her remark. No doubt it would be put down to my foreign travel, but to me it was my love who had changed me on our journey down here with my realisation that she had grown into a lovely young woman.

Now it wasn't our way to rush and tear around. There were plenty of useful and helpful things to do around the home. The outside work in the garden, with both of us clearing and tidying up our spacious grounds, that were far too much for my father to look after. Whilst I dug the potato patch my sweetheart would mow the lawn or attend to the flower beds, then we would break for refreshments, turning the barrow upside down to sit on as I drank my glass of beer. I would remark with emphasis 'Ah, that's better', just like the older men used to say in the harvest field as they quenched their thirst from their bottle of cold tea. My love, sharing the barrow, would sip a pale shandy as she looked at me with a saucy glint in her soft eyes, saying 'Am I going to have a boozing man as a future husband?'

I had not come back having attained high rank or plenty of wealth or been able to say how successful I had been, as I never judged my ability other than normal, but in my humble opinion I felt that all these were nothing compared with the admiration and love I felt for my sweetheart, for often did I think that the Lord had rewarded me, not with brains or the power of command of others, but I was rich in enjoying the humble adventures of life and to me my love for Isobel was far above riches and high rating.

We began to think and talk more seriously of the future. My sweetheart was 21 and I three years older. It was during the holiday that we became aware how deep our love was growing. Life would soon slip by, and we were old enough to make a home of our own. We would talk this over as we walked around the countryside. Often my love would remark that she had put something away for her bottom drawer, with the other things that she had acquired, with an air of secret knowledge which all women seem to have in these matters of providing for the home, implying to us mere menfolk that it was far above our comprehension to understand all that is needed to furnish a home.

The fortnight's holiday passed by. I had taken her back, remaining a day or two before returning home, for we had become so attached that each move by either one of us resulted in careful planning for time off to be spent together. No tearing myself away from the family at the last minute, no rushing to jump on the last train as it pulled out. I was already down there, putting up at the Navy House, just to see my sweetheart for a few hours on two evenings a week. Before my overseas leave came to an end, I strolled back into barracks, not wanting to get another ship, but hoping to become a barrack room stanchion, which was the term given to those who bided a long time in barracks.

Back I strolled, down the long winding road from Chatham to the entrance of the naval barracks. The tall wall of the dockyard rose high above, giving a depressing feeling. It was not until I was in my mess in the large block and talking to one of my old shipmates that the depression of this airy and cold barracks room began to fall away, then with the warm feeling of the grog and the unfolding of the happy and exciting events which had taken place on leave, I began to get back again into the grand Navy life, which seems to have a certain hold over you.

In the next few months life was kind and good to me. After a few weeks in barracks I was drafted to a reserve ship, H.M.S. *Cardiss*, which was lying with others in the basin of Chatham Dockyard, being maintained in working order by a skeleton crew. This reserve fleet was laid up 'in mothballs' to cut down expense, and could easily be mobilised into an efficient force for defence work in an emergency. In my young innocence I didn't give this much thought. It was great to be on a ship which never went to sea. The strict naval discipline was eased and I could get shore leave and long weekends. The summer months were days of romantic bliss when my sweetheart and I would meet, taking long walks along the river bank or through the narrow country lanes.

There were other times on her half-day when we went over to Chatham, taking our seat on the top deck of the bus, enjoying the ride as if it were something special, and there would be tea for two at one of the small cafés. We would get much pleasure from these happy excursions as we talked and made plans for the future. Some Sundays I would get a visitor's pass into the dockyard and take her on board and give her tea, telling her the names of different parts of the ship, as comely a figurehead as ever graced the bowsprit of any seafaring vessel.

The days of summer came and no two lovers could have been more happy than Isobel and I. Her annual fortnight's holiday and my summer leave were arranged without undue trouble. Perhaps at this state of my life the world at large and my country's affairs were hardly noticed by me. Sweet romance and love lifted us out of the everyday worries and the burden of living; with no fear of the dreaded future we only saw life through the eyes of romance, and rightly so, for that summer holiday was the sealing of our courtship. There was no sudden romantic moment slipping the ring on. We ourselves had decided to become engaged.

'I will take you into Cambridge tomorrow and you can pick your ring. That's what lovers do when they are sure of each other.'

'Oh do they?' she had murmured in my ear.

'Yes, I have read it in books, and after a long courtship they are sure they love their sweetheart more than anyone else in the world, and they then take her along to a jeweller's, buy a ring and slip it on her finger.'

'And that means just what?' she asked, as I hesitated.

'That I shall be your devoted slave to love and honour you all the days of my life' I whispered back.

'What do you expect of me for this ring that is going to bind us in a promise for the rest of our life?'

'Just promise to be my wife, that's all.'

'Don't I have to cook, sew, mend, work and care for you in sickness and in health?' she asked me with a glint of teasing in her eyes.

'Not me,' I answered, 'I can do all those things, being a sailor. All you will have to do is to just be yourself.'

'What on?' she quickly answered back, 'just buttons?'

'I will lay my fortune at your feet, just a few pounds, apart from that all I have is health and a will to live,' I replied. This often worried me. Oh for some trade, something which I had a special ability to do, that I could throw my whole heart and soul into. What was I cut out for, just to run along with the crowd to be one of many? I didn't think I was cut out to be a leader, not even on the lower deck. Perhaps when my time in the Navy was finished I could come back to my native village and start something, but for the time being we would have to take each other for better or worse. So one morning during the summer leave, we took the bus into Cambridge and walked around visiting all the jewellers' shops, looking intently through windows at all the dazzling displays of sparkling stones that flashed from rows and rows of golden rings.

'That's the one I like best,' was her final choice as she pointed to one with three stones and into the shop we walked, to be smiled on by an elderly man who gave us the best of service and consideration, which helped to take some of our shyness away.

'I often have young couples like you two. It's a pleasure to serve you. It brings back the memory of my own courting days.' He put us at ease in a friendly way and brought the tray of rings for us to view more closely.

'I prefer this one,' she spoke quietly and pointed to it. The proper size of her finger was taken 'just to make sure there was no hitch at the fatal moment', as the jeweller smilingly remarked, then he delicately placed it in a little box and handed it to me. I paid him and we both happily thanked each other. I put it carefully in my pocket as we departed from the shop into the busy street.

'I mustn't lose it,' I told her, as we walked along, and I stopped to feel it was still in my pocket.

'Let's go in the grounds where those nice flower beds are then have a closer look,' I said to her, 'We have plenty of time.'

It was then, on a summer afternoon, on a seat beneath the tall shady trees, that I asked my sweetheart if she realised all she was giving up to become my promised wife. Heartbreak and heartache, long endless months of waiting and never knowing where I should be, and moreover we had no home. What could we plan or do in my uncertain life? Not until 1940 would my time expire, until then the Navy claimed me. She laid her head on my shoulder as the sunlight streamed through the leafy branches and life went on around. After a minute of thought she spoke. 'I know only that I love you so much that nothing else seems to matter.' I put my hand in my pocket and took out the ring of three sparkling stones and slipped it on her finger.

'Now you are promised to me,' I whispered, and kissed her gently.

'I will try and do all I can to make you happy.' For a minute we lingered, then stirring from my shoulder she stretched out her hand, looking at the ring of gold and sparkling stones on her white finger.

'You don't know what you have let yourself in for,' she exclaimed joyfully.

We caught the bus home and informed my parents that we were now engaged, and Isobel shyly held out her hand, displaying that ring on her third finger for all to see. My sisters insisted that this called for a small celebration and a family gathering. A few friends who happened to come in to our inn were invited to swell the number of our family round the table in our best room on the following day. My betrothed looked beautiful in her summer dress and I wore my sailor's uniform and with family and friends we celebrated our engagement in a nice homely party, with bread and butter, cups of tea, sandwiches and buns, biscuits and cheese and, of course, to make it more proper there were bottles of port and sherry that they all insisted should be opened by my sweetheart and myself. We pulled out the cork to make them go off pop as if it were real champagne. Everyone at the party gave a little cheer and insisted that the sherry and port were excellent, and as if proof that it was, they gathered round the piano and sang to the accompaniment of a gifted pianist and singer. What started off as a small, homely celebration lasted throughout the evening. So many villagers entered, drinking their pints and adding their voices to the old sentimental country songs or swelled to the rousing choruses of the new ones.

Amidst this surprising and happy celebration, we both felt as though a fairy had waved her magic wand and had made everyone gay, and given us a lovely party that no arranging or money could buy.

Throughout the summer and winter I remained attached to the

reserve fleet, just a small number of us to keep the moving parts from rusting and the inside clean and tidy. Most of the crew, if not locals, soon became so or lived close by where every weekend could be taken, which only left two or three of us on board. I was in a position to get whatever night I wanted off, and so the days, weeks and months went by with Isobel and I wrapped in our world of blissful love, meeting and seeing each other on every occasion. There was always the uncertainty of how long this would last at the back of our minds. Then there was the date of our wedding to fix. Often we would talk this over, for we had waited for each other during those two long years of parting. Wasn't that proof that our love was sincere? Had we to let time go by until I would suddenly find myself drafted to another ship on a foreign commission, with no time left to arrange our marriage? And if we did get married, where had we to live? Two years left in the Navy. I didn't want to settle in the Medway towns or stay in the Navy. There was heaps of room at my own home, but I couldn't take my wife-to-be to live amongst strangers, however well she got on with her future in-laws. When on holiday without me, it would be strange and lonely. What had we to do, what could we do? Always these words would ring in my ears which she had spoken when I had told her of the long endless months of waiting, 'I only know I love you so much that nothing else seems to matter'.

'You must fix the date for our wedding,' I said, as we slowly walked back up the long road to the shadowed spot under the wall where we spent many long minutes saying 'Goodnight'. 'For I've heard a strong rumour that I will soon be drafted back to barracks and that will mean another sea-going ship. We must make up our minds that there is some way we can manage, rent a little house or get rooms. Talk it over with your Mum and Dad, then I may get more information of what is happening.' We decided between us that we would go ahead when I had obtained definite information on my transfer into barracks.

'There is a little church right opposite where I work. I would like to get married there,' she told me. 'I watch the happy brides and their grooms come out smiling from the church with friends and relations all gathering round, taking photographs of the two or forming up into family groups.'

'Well, that settles the church. There only remains the date, sweetheart and of course, a place to live. I have a lady who would grace the grandest mansion, but I haven't even a house or living room for her,' I said, as I kissed her 'Goodnight'.

'Don't worry too much, darling. Love will find a way,' she whispered, as I walked back to the ship rather thoughtful and wondering what I could do.

The following week I was officially informed that I was being drafted back into barracks and I would have a day's leave for each month I had served in the reserve fleet, which amounted to ten days. This would be some time in mid-April, the year being then 1938.

CHAPTER TWENTY

I stood before the Commander as the Master of Arms read out the request which I had handed in a few days before, for ten days leave due to 'Reserve Fleet Service', dating from noon on the fourteenth of April.

'Is this all in order?' he asked, turning to one of the Lieutenants standing by who, stepping up, assured the Commander that all was correct by King's Regulations laid down in Admiralty orders, and he understood that Able Seaman Sanderson's request for leave on certain dates was due to his getting married. I stood alert and tense, watching every muscle in the stern face of that officer.

After a good deal of worry and frustration my sweetheart had finally settled the date, for there was a certain amount of discord in her parent's attitude, her father being unwilling to give his good will. Although I had been a frequent visitor to the house for three years and been engaged nearly one year, his unco-operative attitude had resulted in him not speaking, although I was on good terms with the rest of the family. We had rented a room and all was finally arranged. It was up to me now to get permission granted. No wonder I so keenly watched the expression on the Commander's face!

'What day are you getting married?'

'Thursday the fourteenth Sir, at 3 o'clock in the afternoon'.

'Request granted.'

'Shun,' ordered the Master at Arms.

I saluted, about turned, and doubled away. All was fixed, the date set and arrangements settled. It had to be a quiet wedding with my sweetheart's family, a neighbour and a few friends, attending. For me, I was all on my own. No family, not even a shipmate, for there was no let-up on her father's unwillingness. Not a word did he speak to me, making it most unpleasant and I, being self-conscious, didn't want anyone from outside to know. I had written home saying there was no need for any of the family to come as we would be catching the train straight away to spend our honeymoon with them. So it was a silent bridegroom who went about his duty, thoughtful and wondering.

On the very last night of meeting my sweetheart – for tomorrow she would be my wife – we walked to her home, she reminding me that I must be away well before twelve o'clock as it's not right to see each other on your wedding day until you stand together in church.

'You will be there, won't you?' I had asked, perhaps with a catch in my voice.

'Whatever made you ask that?' she answered me.

'I don't know my darling, but I imagine that there may be those who wish you were not marrying me.'

'Don't worry so much. I know it's been rather unpleasant for you. My father may be awkward; maybe he doesn't like the family breaking up, but tomorrow we shall be on our own, making our own way through life.'

She kissed me, then slipped from my arms and was gone down the path. I turned to walk back to those barracks as I had done so many times before, only after tonight it would be all different.

Next morning I set out, walking up and over the Great Lines, the steep white hill that overlooked the town and Medway, with the large War Memorial at its summit, and into the High Street of Gillingham. Turning right, up the streets into Nelson Road, where the Church came into view. I stopped to put on my long white silk ribbons some short distance away from the Church. Now I consciously walked with the white silk ribbons fluttering down to my knees. A few people had gathered at the door and I felt grateful for this as it took away my nervousness. The Churchwarden met me at the door and bade me follow him, showing me to the seat, telling me tht when the bride walked down the aisle I was to tke my place at the altar. I knelt down, and the prayer in my thoughts was for God to make me a good and worthwhile husband. I then waited, not daring to look round as the silent minutes went by. Then the few people present straightened and I stood up as there was a movement close by me and the Churchwarden, by silent look, bade me to take my place by the altar. As I stepped forth, I took a glance to see my Isabella, a white vision, as she put her arm through mine. The vicar began the marriage service; we repeated after him 'I will', and we were proclaimed man and wife before God and the few relations and friends, with the sun shining on the stained glass window, casting a glow of brightness as we knelt down to receive the blessing of the Church. Then into the vestry, where we signed the register. Then we stood in the Church porch while the one amateur photographer 'clicked' and a little girl presented a lucky silver horsehoe. A handful of confetti was thrown by the neighbour and we did a little rush to the open door of the waiting car and away we sped.

We found, on arriving at my new in-laws' home, the table all laid for the wedding feast, a two-tier cake in the middle and the presents laid out around the room. We stood there, hand in hand, looking at the presents with our hearts full of happiness that others could be so good and kind. Our close relations filed back from church and what must they do but open up the French doors, it being a warm, bright day, and have our little party outside on the small veranda. Then the camera clicked again as my bride and I with hands clasped cut the cake, to be toasted with a bottle of wine; all fell to talking so much that we had to catch a later train.

CHAPTER TWENTY-ONE

Back in the barracks, I joined in the company of the natives, who walk out of the main gate every night of their leave to their wives and families. Three nights ashore, then a duty. I walked home to my darling wife. One room in another semi-detached house. It was a start to married life but wasn't there something missing? I felt so but it was so nice and cosy just the two of us for tea and if it was my weekend off, we could go to the pictures on Saturday night, walk back and make a nice cup of drink. Our honeymoon had gone so quickly, to find ourselves behaving so carefully and quietly in this one room. Only the first weekend. Then we were informed where there was a little bungalow to let, all on its own. With this good news we flew around and in no time at all had paid the first week's rent and been given the key. Out buying the lino, furnishings and everything that goes to set up house on our small amount of money, which throughout our lives we had carefully saved for this happy occasion, we ran into my wife's parents in the High Street, and full of excitement told them the good news.

Immediately her father carried on talking to me as though nothing had ever come between us. It was only ten days that we had had our one room. Now we had our own rented bungalow, all cosy and nice, with our new furniture and we were in heaven. There were those nice weekends when together we went shopping, picking out our small joint, I carrying the basket. It was my turn to pay for the weekly trip to the pictures out of my pocket money. There was the kicking out of bed of whoever's turn it was to make the pot of tea, and I mostly lost; the special cooking when we entertained her parents for dinner and tea and the following week her sister and young man; the delight we took in showing them our dining and bedroom suites and what we hoped to do when we could afford it. Gallantly, on each night's leave, I walked in pulling off my service uniform saying to my darling wife that my collar was getting dirty and needed her good attention, remarking how well she washed my white flannels and starched the blue collars. I was the envy of all my service chums. To do my share I bought a pair of shears to cut our small back lawn and I would also show how good a handyman I was about the place, taking hammer and nails and repairing a fallen slat from the wooden fence, hammering in the nail and clenching it down to make it last. Next day we took our tea on the close-clipped lawn.

'You have a look at the fence I repaired,' I said, 'no neighbour's cat can get through. See how firm I've made it.'

'Yes,' she said, as she stood back eyeing it with that quizzical glint in those eyes of hers, 'but why have you put this one slat two inches above the level of the rest of the fence?'

The paradise of married life!

But I had not joined the Royal Navy to roam around in barracks. Three nights ashore out of four, three weekends off before a duty. The Lords of the Admiralty seemed to be unaware that an Able Seaman had got married and was living in an 'ideal haven of living bliss'. How long would this be allowed to go on? Arriving home one day, I took my place at the table laid for two. After kissing the wife and giving her an extra affectionate hug, I told her how nice she kept the house, herself looking so trim and pretty that I regarded her with the undying affection that a sailor has for a slender craft with sails spread, skimming over the water.

'And what has my sailor boy heard to make him talk nonsense?', she remarked. 'I thought you regarded me as one of your "dhobying firms", the many collars and flannels you leave for me to wash.'

'Sit down, my love,' I said to her, and then told her in the gentlest way that her Able Seaman had been called on that day to do his duty, but she was not to get alarmed as the ship I had been drafted to was attached to the Home Fleet, which was never away from its home port longer than three months, adding that I was lucky, and could have been drafted to a foreign service ship and away for three years.

H.M.S. *Southampton* was one of the latest 'Town' class cruisers, of 9,400 tons displacement, and carried a crew of 700 officers and men. It had a speed of 32 knots and its main armament consisted of four 6 in. triple gun turrets, two twin 4 in. anti-aircraft batteries either side, besides two multiple pom-pom and numerous smaller guns, and two sets of triple torpedo tubes port and starboard. It had two large hangers and could carry three Walrus flying-boats to increase its reconnaissance range, which were flown off from a catapult aft of the foremost funnel, with a crane to pick them up off the sea for hoisting inboard. It was a flagship of the second Cruiser Squadron and flew the flag of a Rear-Admiral. Its contingent of Marines, in their coloured uniforms with striped trousers and their smart turnout of guard and band, gave it that class and distinction above the rest of the squadron. This class was kept up even on the lower deck, who boasted that they carried the flag.

I joined H.M.S. *Southampton* with a small number of others. It was lying in one of the basins at Chatham dockyard in the early summer of 1938 and I quickly learned that my first cruise would take me round the Scottish coast. I would see the Forth Bridge and Scapa Flow and view the wrecks of the German fleet, still lying where they had scuttled themselves after their surrender in World War I.

My engagement and marriage had so occupied my attention that I had barely noticed what was taking place in this small but great country of mine. Now that I was in barracks I noticed big headline news in the daily papers. The Germans had suddenly laid claim to various territories and backed it up by taking possession. This had caused a crisis in the diplomatic field, with high-level consultation going on between France and Britain. But while the allies hesitated, Germany acted and the German Chancellor Hitler made big speeches and boasted of the master race of the Fatherland. The British Government called up the Reserves, and the old naval barracks at Chatham, with its figureheads of bygone famous admirals, was suddenly filled by the rush of old hands, who delighted in wearing the blue uniform once again removed from the chest of drawers, where it had been lying in mothballs. They had kissed their wives and patted their children's heads, telling them 'I've only got to show myself and the threat of war will be over.'

How right they were, for after arriving by train, bus and taxi, they drew their rum ration, were fitted up for new uniforms and besieged the canteen in force; some made acquaintance with their old lady friends in the lively, thronging pubs of Chatham. The Nore Command had done its best to cope with these thousands of reserves thrust upon it in an emergency. There was food and clothing, drafting to man the reserve fleet, while gunnery classes were given by the gunnery school to refresh the men and bring them up to date with modern equipment.

I myself was doing my daily task in barracks, trying my best to make myself inconspicuous. As a young married Able Seaman, this was quite natural if I wanted to remain a native, but I was not always able to dodge as some old chum would suddenly run into me saying, with a pretended look of surprise 'You still here? I've been half way round the world again since paying off. You must be the drafting commander's runner or the jaunty [winger].' These reserve hands were very self-assured, and tended to show off to us young and regular servicemen. And well they might, for after a few days Hitler promised to make no more demands, the Fatherland's honour had been restored, the British Cabinet disbanded the reserves, paying them a fortnight's pay, travelling allowance and money. They had all trooped out of the main gate gaily laughing, with their pockets full of money, arriving back to their families with an air of confidence about them, saying 'What did I tell you?'

I had witnessed this and kept quiet, putting it at the back of my mind, for I wondered if this might not be a forerunner, a dummy run for the main call-up which I felt must come. The forewarning of war by the few, who were termed warmongers by those who won the majority of votes by appeasement, pointed out to the people the suffering affects of war.

This was the situation of my country as I saw it when I, a young A.B., joined H.M.S. *Southampton* in the early summer of 1938.

CHAPTER TWENTY-TWO

I had kissed the wife goodbye, saying 'It's out to sea tomorrow,' and now as we left Chatham to sail down the winding Medway, I looked across the river, seeing the many buildings standing along its bank and thinking about our little bungalow. I said in my heart 'Hurry up and bring me back; may the time go quickly.'

Soon the bugler sounded 'Action Stations'. I rushed along to my new gunnery quarters, for I was now the number five on a gun in one of the heavy power-driven armoured turrets. The revolving turret was trained from stop to stop, with the guns being elevated and depressed to their limits. We were all numbered and reported over the phone to the control. 'Y turret closed up and correct', there was a great deal of drilling and practice firing to do to make us into an efficient turret's crew during that summer's cruise.

After all the testing and reporting, the order came through to fall out, and I went down to my mess, seeing the many fresh faces, the new messmates I had to live with, perhaps for the next year or two. Gradually after a week or two, when all had settled in, I had picked out my own personal friend, who I felt I could confide in and accompany on a run ashore. I became pals with a young A.B. who was courting strongly and saving hard to get married. I couldn't go chasing the girls now when we sailed into home ports, we had something in common to bind us together.

That summer cruise passed by without anything unusual taking place, visiting one or two ports around the South coast and writing to my darling wife. Then we steamed back into Chatham and I was back to my little bungalow with the wife all houseproud and ready to practise her cooking on my hungry appetite. There was summer leave to have and the ship would be open to visitors during Navy Week. This was most entertaining, watching the crowds of people swarming up the gangway, walking round viewing the superstructure, the sea boats, anchors, seaplane and catapult, climbing up the many ladders leading to the bridge, then down on to the quarter deck. Y turret was open to allow them to enter and see inside the gunhouse. My pal and I had the task of explaining how the gun turret worked. This we did in such accurate detail that the gunnery officer stepped in, called us aside and said 'Cut it out. You are giving secret information away.'

It was on the winter cruise that we again sailed out from Chatham, heading for the most northerly port of Britain. Desolate, lonely-looking islands lay scattered around as we entered Scapa Flow. There were crofters' homes to be seen, but very few and far apart, with small patches of green which I took to be plots of cultivated land near each homestead. The rest of the low sloping highlands were covered in heather, and around these flowed the sea, where half-submerged wrecks of the once-proud German navy looked bleak and lonely, left to slowly rust away.

This was Scapa Flow, which had seen the might of Britain's glorious fleet assembled in all its strength. The decisive hour had come, in the First World War when the world's two most powerful fleets put to sea. In the battle of Jutland that followed, although the British fleet had lost a number of ships in the early engagements, the Germans failed in their attack on our main battleships and fled back to their ports, never to challenge our fleet again in a general engagement. Their surrender after the war resulted in their scuttled ships lying as monumental wrecks in this inland refuge.

After anchoring, I stood surveying the landscape watching the full drooping clouds lying low over the islands. No wonder I thought, there are so many Scottish people scattered over the world. Where was the enterprise for their standard of education to develop in this almost barren country?

When we were chatting in the mess, one man said, 'It won't be the same again, because Germany hasn't got a fleet, but what they have got, we shall have to keep bottled up. For destroyers it will be all convoy work, going three times the distance by zigzagging, and don't forget a convoy can only go as fast as the slowest one. All around our coast the channel will have to be constantly swept for fear of mines and then there is always the danger of one of their pocket battleships getting through our blockade.'

'The coast of Norway is rugged and full of hiding places, not only for U boats but for large surface ships as well,' spoke up our leading hand, who seemed more of an authority on naval strategy then the rest of us. 'The distance between here and Norway,' he continued, 'is over 200 miles, and to guard against anything escaping by constantly patrolling will keep the whole fleet at sea. Not only have you the bad weather to contend with, but the hours of light get less and less during the winter months. What with the blizzards and high seas running and visibility down to a few hundred yards, it's almost an impossibility. Once a ship escapes into the Atlantic, it's like looking for a needle in one of "Sandy's" haystacks which he used to make on the farm (this was a slight dig at me, being country bred).'

'But we have our old Walrus planes to fly off. They can watch over many square miles of ocean,' I spoke up, 'that is why we carry them, for reconnaissance.'

'Maybe we have,' said the leading hand, taking up my point, 'but after all they are only fair weather birds. Look at the danger of stopping to pick them up, besides the trouble of manoeuvring when the slightest swell is running. It won't be this coming war that the aircraft will be able to take over our duties. The navy will still play the major role, just you wait and see. Take a good look round, for in the near future you are going to see a lot more of Scapa Flow.'

It became strongly rumoured around the ship that the *Southampton*, being a flagship, had been selected to escort the King and Queen on a state visit to Canada and America. There was also an annual cross-country run which my chum Lilley had decided we would take part in to get some runs over the Scottish moors, that being a more sensible way of going ashore for us than hitting the high spots with the lads. Scapa, Invergordon and Rosyth were the sea bases we visited in between our sea exercises and at each opportunity 'Lilley' and I went diligently ashore for our practice run of four to five miles. Every man in the ship had to take part in the ship's marathon and the first twenty would be picked for the ship's team and given daily training to represent the *Southampton* in the home fleet finals at Rosyth at the end of the cruise. Now Lilley could always beat me. We would start off together but soon I would be left behind, to pace along with the slower ones. Not that I was out of wind. I always seemed to be as full of running at the end as when I started.

'Why do you keep pitter pattering on almost the same ground?' asked a runner as I paced along in a struggle to keep up with him.

'Lengthen your stride. That's better,' he said as I immediately put into practice his advice and found to my great satisfaction a big improvement in my running. Next day this same runner couldn't keep up with me. There was only a few days to go before the whole of the ship's company would line up for this event, and as I had taken advantage of every opportunity to train, but had always come in with the slower ones, I had to put up with leg pulling and sarcasm from my messmates.

'With all your physical fitness,' they chuckled, 'we will bet you just a small packet of fags, you don't get in the ship's team.' As I set off on my training, now with my newly discovered improvement, I accepted their bet.

It was a grand sight for the Invergordon housewives and lassies, seeing the spindly white legs, all shapes and sizes of the large numbers of officers and men, lined up in their main street, to start off on the ship's marathon. Even the fat Chef couldn't get excused! The gunnery officer stood on the curbside, franticaly waving us in line and giving his orders in true gunnery school fashion, counting down to zero and firing his Very pistol to send us racing off. Although the fat Chef only went fifty yards before dropping out with the stitch, a large number rounded the first bend and conveniently took the wrong turning, while us others who took it seriously raced along, settling down to get our second wind. I forged along, remembering to keep my lengthened stride going as I began to overtake one after the other. In the last half mile, there were only five in front of me and the next one was my chum Lilley, who had also confidently bet me twenty fags he would beat me. I kept up to him, determined to win that bet. As I struggled along, now gasping for breath, hardly audible were the cheers of the crowd as they stood both sides of the finishing line urging us on. Gathering my last amount of strength, I put in a final burst to sail past him in the last fifty yards and fall exhausted but triumphant over the finishing line. Next day we were both sent for and informed we would receive a medal each for being the first two novices home and would be included in the team which would land daily, if possible, to train for the Home Fleet finals at Rosyth.

The days passed by, with the ship putting out to sea to do practice firing and exercises with others of the squadron. There was always the happy feeling that each day brought one nearer to the return to our home base. My messmates' chaffing had stopped on my excellent performance in the marathon and it was only Lilley who now referred to it, increasing his bet to a hundred cigarettes, declaring he would finish in front of me in the final.

The rumour on the messdeck grew stronger that we were going as escorts to the King and Queen and this rumour proved true when a notice was posted up giving the whole programme. H.M.S. *Southampton* with H.M.S. *Glasgow* would be escorts to their Majesties King George VI and Queen Elizabeth on a royal tour of Canada, Newfoundland and the United States of America. Their Majesties would embark on the *Empress of Australia*, leaving Portsmouth on 6 May 1939. This was exciting news for all on board. At last I would see Canada, that great vast land which the wife had left when a young girl of sixteen, and see those great railway engines, with their homely whistles, that had rumbled their way over the track for three days and nights. America, with its gangsters and film stars, where everything is done in grand style; its skyscrapers; with all its people as equals, no class distinction there. This was my impression of that fast-growing country. I don't think there was anybody on board who didn't inwardly feel excited.

After we had talks on the messdeck of the growing unrest, the rapid rise of the German nation, goose-stepping its way back through Europe, while the caricatures of Hitler were always shown shouting and demanding in gestures, with his Italian Axis partner, Mussolini, boastfully paraded in military uniform, his chest dazzling with rows of medals. Against this, our high government figures at Westminster carried their briefcases, dressed in tail coats and bowler hats, while Churchill puffed harder at his cigar and made his straight-to-the-shoulder speeches against appeasement, saying that a bully was never satisfied and we must rise and arm, time was running out.

The threat of war was now real and I wondered why this should have to come about. There was our small bungalow all cosy and nice. We had added two big leather armchairs. I could now sit with my feet up, reading the paper or a book. Our marriage allowance had gone up from 7s 6d. to 17s 6d., an enormous rise. The wife had written to say she had gone back to work as she had so much spare time. This would help out with the housekeeping budget. We had made plans to buy the bungalow for £320 through the Council, ourselves having to put down the lump sum of £25. At last the tide was turning. What a wonderful feeling to have. Perhaps I would come out of the Navy and find a job round the Medway towns. The rosy future looked full of promise as I swung in my hammock, contented dreams filling my head.

At last Britain was awakening to the fact that her fighting forces had been neglected. The military boot of Hitler was causing some concern to Whitehall. But the announcement of our forthcoming Royal tour did something towards taking our minds off the subject of war. At last we left for Rosyth, the last port before returning to our home base for Christmas leave.

It was a clear, sunny morning as we steamed slowly upstream to see the lattice-work of the Forth Bridge, so clear in the bright sunlight. The ship turned to starboard, towards the dockyard at Rosyth, and tied up alongside the jetty. Night leave was granted for those wishing to visit Edinburgh, but before I could make my mind up on shore leave, an announcement came over the loudspeaker. 'The ship's marathon team will muster on the jetty at 15.30 hours in running kit.' I fell in with the team and we were informed that daily training over the full course would start immediately. In charge was Commander Brock, who seemed bent on getting the best out of us. After a week's training, I felt like a March hare, skipping over the green pastures, up the hills, down in the hollow, through the wood, jumping over the heather banks. We were reaching peak condition, and what made it so entertaining was that we were given a light tea at the sports ground canteen before returning to the ship, at the Commander's own expense we were told.

As the final day grew near, what must my pal Lilley do but get himself persuaded to go ashore to Edinburgh on night leave, and no amount of persuasion would change his mind.

'A night on the beer can undo all the weeks of training,' I said to him. 'why all this sudden shore leave? Has your girl friend upset you?'

'No,' he said, but I didn't believe him.

Our training Commander eased up our hard running in the last three days, and the last afternoon we just walked over the course. His final instructions were to run as a team; if we caught up with a team mate, we were to urge him along by pacing alongside for a short distance. 'I have every confidence in you,' was his final word at our last free tea. The night before the race, my pal and I sat in the mess sewing our given numbers on to the back and front of our vests.

'Don't forget my bet is still on,' Lilley reminded me.

'Don't you want to call it off?' I asked, adding 'you neglected training for a couple of days through going on night leave to Edinburgh. Don't you feel you have let the team down as well as the Commander? After all, he has been good to us, giving our tea each night, which he paid for out of his own pocket.'

'Oh, I know,' he remarked, 'but I am still free, not married with responsibilities which you have. I'm going to have some enjoyment out of life. By next year we may be at war, and my father was killed in the last one. "The war to end all wars",' he said with a touch of bitterness, adding, 'it doesn't look much like it, does it?'

I had no reply to this, trying to understand his feelings. 'I don't think I can beat you this time, but I shall try my utmost,' I said, trying to put an eager note into my voice to lift him up from the morose feeling, asking 'have you written to your young lady telling her of the stiff training you have voluntarily been doing? My own letters to the wife have been full of my physical fitness due to this training we have undertaken.'

We trooped into the dockyard, *Southampton's* band of marathon runners, all in clean kit with numbers to the back and front. There was a general 'make and mend' given to all who wished to watch, for the whole of the Home Fleet had assembled. Battleships lay at anchor in midstream. The Commander-in-Chief and his Staff Officers with their ladies were all present to see some 400 runners form up in teams of twenty, set going in a mass start by the Commander-in-Chief's wife cutting the tape. Then to a throaty roar of encouragement coming from our spectators, we raced up the wide steep incline and in no time at all I could see a good many strung out in front with just a few behind. As for my team, they had vanished. I raced along at a pace I could maintain, passing a good number in the next mile, coming up with a midshipman of our team, pacing along beside him, but not for long. I thought he was holding me back, so with an encouraging word I leapt in front, feeling fine now that I had got my second wind. Where was my chum Lilley? How far I wondered was he in front—for there were runners a long way ahead. I wondered how they had the speed to get so great a lead, but now I began to overhaul one after the other. Our training Commander and several of the team were now in the rear, but as for Lilley, not a glimpse of him so far, and three-quarters of the course must have been covered. I was catching and still passing others but it was now becoming more of a struggle. I gritted my teeth, lengthened my stride, gasping for breath. The last half mile. As I turned a corner a white figure which I couldn't mistake was plodding along, some eighty yards in front. Could I catch my chum? Somehow I quickened, hanging my head, panting in heavy gasps of breath. Although downhill and good going with the dockyard getting nearer and nearer, my legs were so tired and heavy, my drooping arms flailing the air, I was still catching up. Twenty yards now separated us. Still determined in my struggle I could now hear his panting gasps. I had to pass him. Somehow I gathered that extra effort and gradually began to overtake. Then with a noble acknowledgement of 'Go on, Sandy' from him, which helped no end, I panted by, never daring to look back for fear of him passing, but stuck to my task for the next hundred yards to where the crowd waited on either side, the path narrowing as swarming sailors closed in, cheering madly, as I reached the roped-in portion where we could only enter in single file. It was then I looked over my shoulder and there was Lilley only a yard behind me. Somehow I managed a grin, he thumped my back, saying 'Well done'.

Fourteenth and fifteenth we came in out of the whole fleet, fourth and fifth respectively in the ship's team, the ship easily winning the team trophy, a good performance all round. Next day the lower deck was cleared. The plaque which had been received by our training Commander from the Commander-in-Chief's wife was handed over to our ship's Captain and displayed on board as the most valued of sports trophies. We felt honoured receiving our silver medal, Lilley and myself getting an extra one each for the first two novices home in the ship, receiving a hearty ovation from the ship's company. Finally there was the last gathering of the team for a large photo to be taken.

Back home on leave, I talked about our forthcoming cruise. I told my wife, 'On the next cruise you may have cause to be jealous, for I am only human and the flesh is so weak. I'm off to America and the sailors who have been before tell me the rich ranchers come on board ship and offer a large amount of dollars for the lads to desert and work on their ranches with a view to marrying their daughters, just to get English blood in the family.'

We had an ideal Christmas leave together; I had carefully saved my pocket money and my wife's wages were counted out, for there were extras to buy such as the beige-coloured bedspread we had picked out to match the furniture which was delivered by van the next day.

Then back to sea from Chatham to Portsmouth to join up with *Glasgow*. The *Southampton* had been in dock. She was as seaworthy as the skilled hands of British craftsmen could make her for her place of honour as flagship escort, responsible for the safety of their Majesties across the Atlantic.

'Goodbye England' once again, we lay hove to awaiting the *Empress of Australia*. Then she steamed up the fairway, flying the Royal Standard, appearing so steady and gracefully cutting through the water. Her two waiting armoured cruisers slowly proceeded to take up position to port and starboard as she reached the open sea. The note of the bugle sounded from the *Southampton*. The lines of sailors formed up on the cruiser's deck sprang to attention. The Admiral, in his gold-braided uniform standing supreme on the bridge, saluted their Majesties, while the Royal Marine band played the National Anthem. Our royal voyage was under way. The bugler sounded the 'disperse' for all to fall out and carry on their normal sea duties.

As the days and nights passed, the three ships steamed further north. I awoke one morning to see the sea covered in one large mass of small ice floes, the three ships edging their way through, the bows pushing aside the pack ice which had drifted down from the frozen North and was now breaking up as it reached the warmer water of the Gulf Stream. And an enormous iceberg came into view, nine-tenths of its bulk under water. We steered well clear, giving it the right of way. Not for us to put their Majesties' lives in peril, to be caught unawares like the ill-fated *Titanic*. Now on the distant horizon we saw that we were nearing the entrance to the St Lawrence. Although I'd seen the Atlantic rollers and iceberg amidst the ice floes, my main interest was to see Canada, the birthplace of my wife, of which she could never speak without that faraway look coming into her hazel eyes. I had told her that I would take her back on a holiday some day in the future, so that she could go back down that long railway track.

I stood gazing across at the land we were approaching as we sailed up the St Lawrence, the forest of fir trees stretching far inland, just as I had thought it to be, vast, lonely and wild. At last Canada stretched before me and tomorrow I would set foot on it. Quebec, where General Wolfe climbed the steep heights and beat the French in battle.

The King and Queen disembarked from the liner, the two cruisers laying off while this ceremony was taking place. Half of our trusted mission was over, as their Majesties were to travel overland. The two cruisers were to spend two days here then on to New York, Boston and finally Halifax, Nova Scotia where we would rejoin the *Empress of Australia*, taking up our royal duties as escorts on the journey home.

On going ashore and setting foot on this northern continent for the first time, I went first to see those railway engines which pull the trains 3,000 miles across the country from the Atlantic to the Pacific. Then into the town of Quebec. My impression of these French Canadian people was that they still regarded themselves as French and would pick a quarrel to fight and carry on the battle that lost them the country. Should you forgive and forget or carry on fostering hatred in your hearts for centuries? This is what the Catholic French seemed to be doing. Thank goodness our British democratic way leaves it for them to decide.

Then it was on to New York, the city of massive skyscrapers, and we tied up alongside the jetty opposite 51st Street. Now I had to go ashore and wander around New York, and I found it impossible to get lost, its numbered streets running one way and numbered avenues another, but like London, you can become very lonely in a crowded city, and it was a foot-weary sailor who walked back to his ship after trudging around blocks of New York buildings. From the dullness of the first night to the brightness of the next, when the ship gave a ball at one of the large halls near Times Square, with the Royal Marine band in their smart brilliant uniforms, giving to British New Yorkers a moving touch of historic Royal England that warmed the hearts of our British patriots.

I myself scrawled across a piece of paper in lipstick an invitation to go riding with some young lady whose father, she said, owned a ranch. Next day as I walked through the docks I wondered whether to keep this date; if I could ride a horse accompanying a young lady on gallops across the countryside. Having my doubts whether I would make a cowboy, it didn't take long to change my mind when a shipmate asked me to go along with him. There followed hectic times of drinking and eating with these New Yorkers, whose parties never seem to stop.

In the evening at the Bronx, a suburb of the city, a small party of us travelled by subway train, which I thought was much below the standard of the London underground railway. There, tucked away in amongst some leafy trees was a country clubhouse where we were entertained in a warm and homely way by our overseas country people, mostly Scottish, who had laid on tea and talked to us of their homes they had left in the British Isles. The dance band in the corner of their small hall, with the ladies sitting round, reminded me of the dances at the village hall. There were steps leading down to the cellar where their husbands popped for a wee tot of their Scotch whisky. It wasn't until the early hours of the morning that their sons and daughters piled us in their cars to return us to our ship.

I got myself ready to go ashore next day, thinking that I would stop for a drink in Jack Dempsey's bar. It was in the street right opposite the jetty we tied up to. Setting off, walking briskly over the gangway making my way along the dock in the brightness of the afternoon, I hadn't got past the ship when confronting me in a hesitating manner was a young lady—and the loveliest, was my observation, as I stopped in my tracks. Surely some officer's friend? Perhaps the admiral's. Not for me to put myself forward, so I went back a step.

'Can I help you?' I asked, and I was aware of hearing the sound of my own voice.

'I've come to see over your British ship.'

'Have you arranged with some officer to take you over?', I asked.

'No,' she replied, 'I am all on my own,' saying this as if she was taking a bold step and was slightly scared at being so forward. Perhaps our life as sailors, mixing with the worst types, makes us sense the bad types, whatever their graces, but confronting me now, speaking so softly, was a young and pretty girl, innocent and truthful as ever I had met in any country, my own or anywhere in the world, and this was New York, known for its crazy-living people.

'Can I show you over?' I asked, still aware of hearing my own voice.

'It would be most kind of you,' she said, adding, 'don't let me delay you from wherever you were going.'

'No, I wasn't going anywhere. I have nowhere to go,' I hastily told her, being a bit untruthful but if I did have before, I hadn't now, for everything was gone from my head apart from this lovely girl standing before me. I bade her walk along with me.

'Would you like me to make you a cup of English tea?'

'I would love to taste your English cup of tea,' she answered.

'If you follow me, I will show you down to my mess,' which was two decks below. I had the job of attending to the two regulating officers who had a self-contained mess. Being very attentive to my pretty guest, seeing that she didn't bump her head on the hatch way, holding the curtains to one side, I conducted her to the small private mess which could boast of a little table with a white tablecloth and two chairs. I was trying my best to spread the butter on the bread and cut it thinly, when a slender hand reached out and she asked if she could assist, so I attended to making the tea. When all was ready, we sat down on the two chairs.

'These slices of bread and butter are almost as dainty as the hand that cut them,' was my remark.

'And your tea is as nice as the sailor who made it,' she coyly replied.

'You will be able to tell others you had tea on board a British warship.'

'Yes,' she answered, 'Mummy and Daddy would think I had gone mad, coming all by myself to the docks to look over a warship.'

'They need have no fear,' I told her. 'How did you know of the ship and what really made you visit us?'

'I read in the paper of your visit and I wanted to meet some English sailors, those daring heroes whom I have read about.'

'Have you been in America long?' I asked.

'My parents come from Scandinavia,' she told me. Now she was smiling, her head turning with the flaxen waves of gold sweeping round that lovely white neck, as she spoke of the rash deed she had so brazenly taken, the courage to come all by herself. Not even her mummy and daddy knew. 'The devil must have got in me this afternoon,' she laughed, filled to the brim with excitement at her adventure which had turned out so wonderfully successful. 'More than I ever dreamed of,' she sat telling me and she asked me to repeat 'Yes' and 'Thankyou' in my English accent, until looking at her watch she said she must go or she would be missed. 'Can I come again, and have tea with you on your ship?'

'Yes,' I told her, 'but we have only a few days of our visit left.'

'I will come tomorrow about the same time,' she replied, rising to put on her light summer coat and looking in our mirror to tuck in those flaxen waves under her broad-brimmed hat. I dutifully escorted her off the ship along to the street where we shook hands, she thanking me for the wonderful entertainment and assuring me that she would return tomorrow, hurrying off to be swept from my sight on the crowded pavement.

The following day when the ship had been cleaned up, grog been issued, dinner laid on, perhaps not eaten in the haste to get ashore, I stood and watched from the vantage point of the fo'c's'le deck, not long for the thinning out. Very soon it had dwindled down to a stray one or two that were taking their time in the hot sunshine. I was also watching out for a wide-brimmed white hat coming from the street. I had laid the table, bought some cream cake, biscuits and a small tin of fruit, also some white sugar. I had polished the mess-shelf and all was shipshape for my pretty guest. I was ready long before her expected time, knowing she would come, but it was entertaining watching others going their way. The more who went ashore the better. Much nicer to entertain one's guests when the ship was almost deserted. Then, walking along in the hot afternoon sun, I caught sight of my girl friend. I was off to meet her, putting to rest the fear she may have that I wouldn't be there.

'I've been watching for you,' was my greeting.

'I hurried off the moment I could get away,' she answered, her blue eyes sparkling with a pleasing smile as I escorted her over the gangway, seeing the quarter master and marine sentry partly taken back. On looking over my shoulder both were staring in our direction (I guessed they were saying 'how does a half wit with a mug like his get hold of a lovely bird like her?'). We went down to the mess. With a little touch of modest homeliness on her second visit, my visitor put herself at ease by straightening her hair in our little mirror as I hung up her coat and hat. Then she turned to the mess-shelf, bringing out the small tablecloth, laying the table with enthusiasm adding a feminine touch of comfort to a man's world, with me telling her that fate sent her, not only to put my mess in order, but to keep me straight as well, for I hadn't been to any gay parties, but had spent the time preparing for her visit. She laughed and bade me 'Sit down, while I make you a cup of your English tea'. I could only sit and watch with admiration, enjoying being the guest in my own mess, at the delight she showed when placing before me the English cup of tea, 'made by me,' she exclaimed with a little laugh at her achievement. It was then we fell to talking, she wanting to know all about England, my life and parents.

I became a little uneasy at the rapturous glow that showed in her face, and the intense feeling put in her voice. For I held back from her that I had wooed and won and been married these last twelve months, but her little outburst of delight stopped me from saying any more. I didn't think my simple story would have such a moving effect. I had better change the story or get going or I felt she might throw her arms around me in her affectionate mood.

'Won't you show me something of your city? I only see the fringe of the bustling cities and ports. I long to go into the heart of the land, to meet its people.'

We finished our tea. She would show me something of New York. We left the ship, walking along into the city. But where was my lovely girl friend taking me? I had been through the doors of dance halls, speakeasies, bars, drinking dens, all the high spots that sailors boast about, but never anything like these wrought iron gates and Gothic doors. I looked up, to behold angels in stained glass windows, for it was the Church of our Lady, and facing down the aisle, my fair companion crossed herself, kneeling down, bowing her head in prayer before the figure of the Virgin Mary, cradling the holy baby, the Son of God. She plucked at my sleeve for me to follow to a pew, both kneeling for a minute. She turned her head whispering, 'I see you are not of this Church.'

'No, I'm Church of England.'

I noticed a confident smile hover over her face at this as if thinking 'you would convert yourself for me'. She told me the Church was open day and night. There was always a priest present. I kept silent, and we soon left, to walk through Central Park, taking a seat on a park bench to watch the red squirrels playing. My fair companion seemed thrilled with her afternoon outing, regretting the time passing, leading me back to the riverside and finally declaring she must hurry off, turning to wave her little hand in sheer delight, without making any arrangements for another meeting, but I suspected that this was not the end of our friendship.

Next morning, reading my name and address in neat handwriting, I wonderingly picked up the following letter.

'My Dear Friend,

Thank you very much for your friendship. It has been so excitingly wonderful being with you, and having my teas on board your ship. I shall never forget our walk through the park. Both my visits will always remain treasured memories to me. I do hope

this won't be the last of our friendship, but only the beginning, as the last two days have been the most wonderful in my life.'
   Yours sincerely,
(With her name and address both at the top and bottom of the page.)
   An invitation to resume our acquaintance! I shook my head. Not for me, for I saw a picture of my wife. She was still my sweetheart, the only one I loved. I looked down at the letter lying on the table. No harm or wrong had been done. What sort of sailor was there that wouldn't have enjoyed the companionship of this young girl? I had treated her sincerely and respectfully. I screwed the letter up. Far better to let the young lady's two most wonderful days in her life remain as such. If I had been single it might have had a different ending.
   Just a glimpse of the Statue of Liberty appearing through the mist-shrouded bay on leaving New York for Boston, Massachusetts. Once we were in Boston, suddenly turning a corner and looking up I saw a sign pointing the way to Cambridge, my home town in America, with its University of Harvard. This was also the birthplace of the colonisation of America. I was shown the first rock the Pilgrim Fathers set foot on when they landed from the *Mayflower* in the year 1620, the graveyard in which they lay buried, the privation of their voyage with the hard winter and hostile Indians having resulted in half their number dying in the first year.
   From Boston to Halifax, Nova Scotia, Their Majesties paying the last call of their tour with a final inspection of the two escorting cruisers. It was a chance for me to see a close-up of my King and Queen. There was the whole ship's company lined up in rank and file, the Master at Arms stretching his fat body to look its tallest, leading the piping quarter masters, the shrill note announcing the approaching royal barge. As it drew alongside, the Royal Standard was run up to fly from the masthead. The click of heels, the flashing of bayonets as the Royal Marines presented arms, the glitter of gold from ceremonial swords and uniforms, of the ranks of officers standing stiffly to attention on the quarterdeck, with rows of sailors lined up in file on every deck. Not a sound, not a stir. Only the puffed cheeks of the grog-faced Master at Arms became inflated and the drawn-out note of the bosun's pipe died in the silence. The Admiral advanced to present his officers and ratings to the King. I stood

awe-inspired with just the pupils of my eyes moving as the King and Queen moved along our file with not more than two feet separating us. I felt sorry for the worn face, the pouching bags under the eyes of His Majesty, but making up for all that the King lacked, followed his stately, elegant, beautiful Queen, whose grace and charms enslaved the hearts of us all. After their departure and the welcome signal was received 'Splice the main brace', we now added to the toast 'God Save the King, preserve our Queen'.

Sailing home, and arriving at my home base, I made my way home to our little bungalow. Not many more months to go now I told the wife. Should I find a job? I had seen something of the world. But nothing was decided. It didn't trouble us all that much, our life was full of married bliss.

CHAPTER TWENTY-THREE

Those last few weeks of summer, lying in Chatham dockyard, when the threat of war was growing stronger and causing concern to the ship's company of H.M.S. *Southampton*, all those who were bona fide natives slipped home on every available occasion. On returning from leave in the morning, we purchased the daily newspaper and viewed the headlines telling of German forces goosestepping their way through Europe.

We did not underestimate our enemy, but had faith in our own understanding of the sea, to adapt ourselves whatever the conditions, whatever the force encountered. We never thought of running away; if a destroyer encountered a German battleship it was 'Go in and ram!' Our seafaring skill still reigned supreme in our Navy from the Admiral down to the boy seaman. I said very little, but listened and wondered what others were feeling. Scapa Flow was freely talked of, and when an announcement came of leave to midnight, nobody to leave the Medway towns, with one voice we all exclaimed 'It's Scapa for sure.'

I walked home to our little bungalow for the few hours of freedom, saying to the wife 'I don't know if this will be the last time, for we are at short notice for sailing and all the lads say it's Scapa for sure'. We had talked this over if war should come, but still not knowing what to do. We could pack up our bungalow, the wife going to live with her people as one family, for the dread of air raids was talked of and the Chatham dockyard was given number one priority as a bombing target. And then there was the little infant we were expecting, the uncertainty of what would happen. What best to do? I could only advise the wife to do what she thought best, for we never expected the war to end quickly. We never thought of losing, we thought of it as something dreadful hanging over us. So it was with a heavy heart I watched the hands of the clock ticking round to midnight. Then I kissed the wife goodbye to walk back, wondering what life held in store.

Next morning we glided out into the Medway. I wondered as we swept by Gillingham Strand when next I would sail home. Then the morning mist hid from view the local riverside I had come to know so well. We swarmed over the ship in a rush when 'action stations' was sounded, for our secret destination would be announced by the Captain when we were all closed up, or the reason for our hurried departure. When the mustering and reporting was over, we waited at our quarters. Then the words 'Stand by, the Captain will address the ship's company'. A calm settled over the group of men gathered round their guns, the quietness, the tenseness of expectancy, the crackling of the loudspeaker, then the distinctive voice of the Captain.

'You are all aware of the crisis. The consultations that are going on between heads of states mean that the Navy has to be ready in a strategic position to deal with any situation that may develop from day to day. We are now on our way to take up that position. This is all I can tell you.' The crackling of the loudspeaker resumed, but it was soon drowned by the hushed tones and tense voices.

'It's Scapa for certain. You bet the French and our forces have already started firing on the Germans.' Slowly it dawned on me I had volunteered, been trained to defend my country. Now I was about to be called on to fulfil that obligation. In another six months I would have been a free man. We steamed on and the following day we dropped anchor between the low-lying islands of Scapa Flow, with the German wrecks of the First World War as reminders of what might happen to us if we lapsed in our watchfulness.

While we took up our strategic position, the gathering war clouds grew blacker. Hitler, now drunk with power and brushing all reasonable proposals aside, moved in to crush Poland, while Britain and France solemnly reviewed the situation. In the last few days of aggression by the German leader, Britain's silent Navy assembled in its wartime base, and in the last week of August 1939 H.M.S. *Southampton*, with other warships of the fleet, put to sea. Our job of bottling up the German ships was now to be put into effect.

We were at sea, ready and waiting, guarding England as our seafaring fathers had done throughout the centuries. At last Britain and France were facing up to their guarantee of defending Poland if attacked. Their final note to the German Government stated 'Unless they withdrew their forces by eleven o'clock on the third of September, 1939, a state of war existed between us.'

'Paint all over the brass work' was the order of our Petty Officer. No order could have been carried out with greater alacrity. It was a general rush to the paint shop, a scramble to reach the upper deck to dab grey paint on pieces of brass that had meant in the past hours of polishing. We now knew there was to be no more practising at war—it would be real. We were to remain by our guns, waiting for tomorrow when the time limit expired, guarding the sea. But what was that overhead? The Graf Zeppelin was flying down the East coast, well within the three-mile limit. Obviously it was taking aerial photos of our East coast. We could not bring it down, however—that would be an act of aggression. We must wait for the final hour.

On that Sunday morning of 3 September we were gathered in our armoured turret, with the rear door opened. The final hour drew near and our turret officer brought in his portable wireless so we could hear the Prime Minister broadcast. I looked around in this hour of crisis which would mean perhaps death to some of us, wondering how many of us would survive; how many would give their lives in this war about to be declared. Fifteen young men, good, honest, hardworking, keen and adventurous, the pride of the country, with our lieutenant, fresh-faced and handsome. No cut-throat crew of swashbuckling villains, ready to kill and plunder, but ordinary men, thinking not of winning and gaining territories but of defending their country, their homes, and the lives of their children, wives and parents. Slowly the minutes dragged by. Then the voice of Neville Chamberlain came from the square box that all eyes were riveted on. 'This morning, I have to tell you no such undertaking has been received and a state of war exists between us.' We looked at each other. This is it.

'All guns load with HE shells,' came the order from our communicating number. In a few seconds from the alarm being given, the whole weight of our explosive shells could be sent hurling through the air, in their flight of destruction, wrecking and sinking, killing and wounding. We too could be blown sky high in a few seconds. Mines, submarines, aircraft and the enemy raider were to be watched for, spotted and reported in a matter of seconds. As if to remind us of the deadly peril, the vibration of the ship increased as she slewed over, zigzagging in case of the lurking enemy. I thought of the wife at home. Already she might be in the midst of air raids. I hoped she would go to my village, to the safety of the countryside.

This was to be my home, or prison, cramped in this shell of iron, eating and sleeping by the breech of the 6″ gun, ready at a moment's notice day and night to load and reload; our lives, our ship, and our country looked up and depended on us. I thought of the lines 'Oh God, our Navy we adore, when in danger and not before'. We would be heroes.

I fell to thinking, now that my life was at stake, what life really meant. Weren't we considered almost as serfs? 'What school did you attend?' was asked when you joined. 'Our village school,' we spoke with pride. This was promptly written down and held against us ever afterwards as a third-rate education. 'High School,' one recruit emphatically declared.

'High School, young man?' the interviewing officer enquired, looking up.

'Yes Sir,' did the raw recruit declare. 'It was built right on top of a hill.'

Then there was the time I couldn't get a third-class seat on the crowded train, and shamefaced got into a first-class carriage and quietly sank back into cushions which I hoped would hide me. After a long time I had the audacity to ask one gentleman if I could have a look at his paper which had been lying untouched by his side. He turned and raised his eyebrows as if noticing me for the first time, remarking with a haughty voice that third-class carriages should be available for third-class passengers. I never forgot the wound of indignity.

I thought of my village, the life of its people, how we had to bow and scrape to the highborn, touching our hats and bowing when the Squire went by. 'You must touch your hat and call him Sire,' said the old hands as I worked as a boy in the fields. 'Why Sire and not Sir?', I once asked, to be told, which puzzled me at the time, that he was Sire to half the village children, for if they got threatened with the sack from the estate, it was generally the man's wife who visited the squire in his moss-covered mansion and got him reinstated. But for all the bad things, my country was worth fighting for, and if we won a lot of these bad things would disappear.

I wondered what the others were thinking as we steamed on our zigzag course with guns loaded. The next few hours sped by as we sat around, not caring to divulge our thoughts or intervene in the minds of our chums. But now there was some development as the communicating number suddenly sang out 'Stand by'. Every man was alert and ready as if an alarm had been given. The speed and vibration increased. The turret officer sprang to look through his periscope as the large turret slowly swung round. Was it a raider, aircraft, an enemy force we had run into? Then the turret officer reported there was a merchant ship ahead. We stood at our posts but released our grip as we wryly grinned at each other at the first standby order given in war. We steamed close by the merchant ship, which appeared stopped and settling down while her crew prepared to take to the boats. 'They have opened the seacocks and are sinking her' was our remark as we swept past. Two destroyers from our force were ordered to stand by. 'A small setback to Hitler' was our comment on this first day of war.

On the twelfth of the month we dropped anchor off Sheerness; nine days of patrolling, a good breaking-in for the ship's company. We had as yet to fire an angry shot. The sea time and practice warning were getting us keyed up for that vital moment. Now in the well-fortified port of Sheerness would we get some leave? Already the Londoners were checking up the times the trains would take and the many natives were listening for the loudspeaker to announce what leave would be given. Then there would be the mail to pick up, with all the news of how each one's family had fared in this upheaval of their everyday life.

It was a busy time. There was the Oiler to secure alongside, and the smaller tugs bringing in the provisions, with all the spares and necessary equipment to keep us going, for now there was no rest, not even in harbour. Men had to remain close up by their guns. Eagerly and hopefully I listened to every announcement. At last it came. Leave for part of the watch from 1300 to 1600 hours. Not much time for me to race to Gillingham, twenty miles down the line. Not worth the risk of breaking leave, not in wartime.

'Sanderson,' a voice hastily called out.

'That's me,' I replied, springing up.

'Report at the Master at Arms Office.'

In two strides I was up the ladder and once more I was at the small office.

'Able Seaman Sanderson, told to report here,' I said.

'Telegram for you,' he said, handing me the yellow envelope which I took and turned away, eyeing it as I now slowly and wonderingly opened it, to take out a slip of paper and read 'Son, 6½ lbs. Eleventh. Both Well.' I wandered along to the mess and sat down, the war forgotten. I was now a father. I jumped up. I must go home to see my son, swift news called for swift action. I suppose I should make out a request–but in no time at all I was standing before my Divisional Officer, requesting permission to travel to Gillingham to see my wife and newborn son, with the telegram as evidence that I was now a proud father. Lieutenant Layton patiently listened to my request and then wandered off. I remained, hopefully catching my breath in the long, tense minutes that followed. Returning, he motioned me towards him. Then speaking confidentially, and not like an Officer addressing a rating, he said 'Can you be sure you can get back on board by 1600 hours?'

'Yes Sir,' I replied.

'Well, no one can give you leave to leave the town, but I know we can't sail until that time, so it's up to you. Now don't let me down. Good luck.'

I clicked my heels, saluted and was gone, with untold thanks remaining for ever in my heart. Ashore and on Sheerness station the train rattled down the line and I noticed a few of my other shipmates were doing the same.

'What time would a train leave to get me back to Sheerness by four o'clock?' I asked. Almost an hour at home I thought, as I hurried from Gillingham Station reckoning up the railway inspector's information. Now I ran through the streets and back paths to my mother-in-law's home and burst in on that unsuspecting good soul, whose eyes lit up with surprise, remarking 'Well, I never'. Then she bade me wait as she entered the other room and I heard her say 'You have a surprise visitor.' Coming out, she asked me into the room and closed the door. There was my dear wife, wondering who. Then her eyes glistened over with love and joy as she beheld her sailor boy, and I knelt down beside the bed, enfolding my dear love in my arms. Then she placed her hand for me to keep silent, gently parted a soft white shawl from a small bundle by her side. 'Our Son.' I saw a red little face with a tiny pug nose, little eyes tightly closed, a wee arm, with ever so tiny fingers, which I so carefully touched. Then my dear wife turned and watched that small bundle and I was aware of a mother's love for her child as a look of melted tenderness stole over her face. 'I have to share you now,' I said, turning to her.

'You gave him to me,' she whispered, 'so I shall love you both.'

'I could have screamed in anger the first day of war,' she told me. 'The sirens went fifteen minutes after war was declared and all rushed for their shelter.'

'We haven't fired a shot yet,' I informed her. 'One merchant vessel sank itself at the sight of us, so don't worry too much about me. I only have 40 minutes and I've given my word I will return on board by 4 o'clock, so your wartime sailor must soon be away. I have a much more precious and larger family to fight for and will be for ever watchful so nothing gets through to harm you. Perhaps the war won't last very long and we shall be together again for always. Look after him and take great care,' I said as I gazed at our little bundle of heaven, kicking and stirring in his innocent baby sleepiness. Then he slightly opened his sleepy eyes and I whispered that he was having a peep at his father before leaving. I kissed my wife, saying 'I'll be seeing you', turned away and was gone from home, saying cheerio to her mother, who followed me down the garden path to the back gate, and I hurried away along the road to catch my train, heavy of heart at having to leave but happy and thankful at having been able to see my wife and son at this vital time. I caught the boat out to the ship at 1600 hours, returning without incident, the *Southampton* leaving Sheerness soon after to take up patrolling duties in the North Sea.

CHAPTER TWENTY-FOUR

There are many Admiralty laws and orders for servicemen under wartime conditions; with the safety of the ship uppermost, it's ship's company are well aware of their duty, but the advances of speed, weapons, and forms of attack put out of date King's regulations and Admiralty instructions. No longer was a Man of War secure from attack even in harbour, 200 miles from the nearest enemy base. Aircraft reconnoitring was taking place all along the East coast; a system of green, yellow and an imminent red warning that enemy planes were in the vicinity meant that men had to rush to their post, perhaps just after entering harbour. I am sure if Hitler had only been aware of the conditions that existed in the first few weeks, his airmen could have come over, flying at leisure, and placed a bomb down the funnel of each anchored warship! Even our 6″ turrets were locked up in harbour and written permission had to be obtained before the key was turned over. With all this red tape and a dozen of these red warnings a day, impolite words were expressed rather freely, especially when all the raids proved negative. Perhaps a certain amount of red tape is necessary and tradition dies hard in His Majesty's Forces, but usually one has to learn the hard way before changes are made by the higher authorities, which are so arduous to the men of the lower rank.

We were steaming in force one dark October night. Very soon a rumour spread round that we were a covering force for a submarine which had been crippled by enemy action, unable to dive, and making slow progress on the surface; our plan was to attract the enemy away from the crippled target. Our four cruisers with an escort of destroyers were a formidable force to encounter, and no doubt there were other ships in the vicinity which we knew nothing about.

Early next morning our communicating number reported that we were being shadowed by enemy aircraft. What did that mean we wondered—but not daring to ask out loud? Would they send a surface force against us? Now we were having a taste of what war was really like. What was that? The crack of the 4" gun sounded in the silent shut-in turret. They must be trying to bring the shadowing aircraft down, I thought, giving a little ease to my nervous tension. But then a deeper boom sounded through the rapid crack of the 4" guns, followed by the thump, thump, thump. Underwater explosions.

The thought jumped into my mind that we were under fire. Then came the information that our force was being attacked by high-level enemy bombers. Thank goodness, I thought, we were four cruisers screened by destroyers. Those 4" guns would be able to keep the planes at a respectable height to make it difficult for accuracy. I breathed more freely again as the many cracks of our guns continued, with now and again the deeper thump of the more powerful bomb. Its vibrating shudder was our only way of telling or judging the distance away.

The first baptism of enemy fire. No doubt, I thought, they are trying out the accuracy of the bomber against the guns of the warship, as the raiders continued raining down bombs, with shell bursts dotting the sky and neither side appearing to get any advantage. It became rather boring to us in the 6" turret, not being able to do anything or see much of what was going on. Some of our numbers had to be sent down to the 4" magazine to assist the hard-working magazine crew, for the planes kept attacking throughout the day and only tailed off when the gathering dusk put an end to the battle. No doubt the high-ups on both sides would work out the effects of the bomber against the warship. We had no casualties. A few bombs dropped close and our 4" magazines were depleted of stock, but this was soon rectified when he had raced for home, dropping anchor in the morning, not far from the Forth Bridge; ammunition tenders soon came alongside to replenish our diminished stock.

The weary ship's company were working hard to be 100 per cent efficient again, ready for the next encounter with the enemy. When all had been made good and our views had been aired, as we related our experiences under the attack, a 'make and mend' was given. No hands required—only those unfortunate ones who were duty watch.

The washing up was done and all was shipshape. I gathered my pillow, placed it on the mess stool, laying my weary body down to rest, for already loud snores came from sleeping sailors, whose bodies cluttered the messdeck, lying on coats spread over the deck or stretched out on stools and tables. Jack was asleep, weary from his brush with the Hun, safe now in harbour, well up river, where we were protected by a boom across the entrance. There had been a few red warnings in harbour, but nothing ever came of them. Anyway, the Royal Air Force could deal with anything that dared to approach inland. And so, with others, I was soon in the land of dreams. There would be a letter to write home as soon as I awoke.

Startled, I sprang to my feet off the messdeck, thrown from the upturned stool by the shuddering ship as a shattering explosion vibrated through the whole framework of its construction. I raced up the ladder. No stopping to think out what it was, for all day yesterday they had dropped around us and now the boom was inprinted in our hearing. Tearing along the upper deck, dodging and ducking from the flying figures of others closing up at their action stations, reaching the turret only to find it locked up. A whole bunch of us fumed for the turret sweeper for the key, when suddenly someone shouted 'Look out'. A roaring wind filled my ears and staring up I saw a plane with black crosses turning into a steep dive, heading down straight for us. I took a few seconds to take this in and then I sprang for cover, but lo and behold, those few seconds cost me a sheltered place under the turret. Fat backsides stuck out from underneath and pull and shove as I did, I couldn't shift the human wedged-in bank. With the fierce whining bursting in my ears, I jumped for the only gap I could see. That was the quarterdeck gangway. I crouched on the steps as the shattering thump of the exploding bombs rocked and shook our anchored ship, heaving the quarterdeck gangway as I hung on to the lanyards, quickly springing on to the quarterdeck again as I realised it must have been a miss. But no sooner did I leap up than the whining, droning pitch of another—ratings jumping into their shelter under the turret and myself jumping down the gangway. Again the shattering explosion, and I realised I was still safe, leaping again on to the quarterdeck, only to leap back in a hurry, to crouch low down as the growing, terrifying wind came and the rat-a-tat and pinging machine-gun bullets sprayed the water close by my feet. Glancing up I could see the plunging plane in its dive coming straight for me. I clutched the lanyard and gangway in a grip of fear for I felt sure the plane must bury itself in its target. Through the terrifying din I heard our 4" open fire and its crack was like sweet music to my ears. The large silver body of the plane soared over the funnel tops, then shattering debris flew around, bits of boats descended and dropped by me, as I slipped into the drink, still maintaining my grip on the gangway as bits and pieces went floating past. I scrambled back on to the gangway platform, noticing the motor-boat tied to the ship's boom a moment before had been blown to smithereens and the swift current was carrying the debris away. Luckily for me, I hadn't let go and I rushed up the gangway on to the deck again. Now somebody had found the key and unlocked the turret. Jumping in, we rushed to our action positions, not that we could do much with our big guns, but no sooner were we assembled than I thought we were heading for the bottom. The stern suddenly rose straight up just as if she was taking a dive and I thought 'Damn, I was safer outside!' as I saw the difficulty of us all trying to get out of the turret if it went down. It righted itself, however, and I breathed again, as I realised I was still all right.

The communicating number now manning his headphones called out 'Control to Y Turret. Are you all right?' 'Y Turret closed up and correct,' was reported back, to be told a stick of bombs had just burst under the stern. To think I had just got inside in time. We heard one or two only further away, waiting with realisation slowly coming back of our tension in the hectic last ten minutes, beginning to talk amongst ourselves, wondering what damage had been done and whether anyone had been killed. It hadn't been so bad being bombed at sea, where all were prepared and ready, but to get caught napping in harbour was different. A great deal of thinking would have to be done and speed of action must result from this enemy operation. For no longer would we feel safe in the sheltered waters of our inland rivers and bays. No more turrets locked up when life is at stake! Speed of action is required in this modern war.

Now allowed to fall out from the turret, with the smaller defensive armament remaining at the ready, I trooped back to my messdeck to view the damage. The bomb had struck the bridge. Jerry couldn't have had a more direct hit. The bomb, being armour piercing and falling at an angle, had gone through each deck, making a wider hole at each deckhead, on through my messdeck, and finally had gone through the ship's side, leaving a hole a yard wide, luckily above the waterline, to explode in the water under the boats moored to the ship's boom.

Alas, there was one casualty: a flying bit of metal had hit the head of one boy seaman, fatally wounding him and bringing home to us the sorrow that all must bear, especially the mothers, fathers and family in wars.

Altogether, we were told, seven dive bombers had singled us out for their target. Being almost unprepared and a sitting duck, we considered we were most lucky to come out of it so lightly. No doubt the Nazi pilots hadn't realised they had hardly any opposition. Later we heard on the news that an ineffective attack was launched by the German warplanes upon British warships in the Firth of Forth, carried out by twelve to fourteen dive bombers. The cruiser *Southampton* had been slightly damaged. The German planes had been chased away by R.A.F. fighters. Four enemy planes had been brought down.

We viewed the damage to our ship. It would mean the dockyard. Would we get 72 hours' leave? I could be rattling down the line on the Flying Scotsman to describe my experiences. Alas! within two hours we were patched up, ready for sea, rapidly learning what total war really meant.

Already the U-boats had sunk an aircraft carrier and now as we sailed up to Scapa Flow, we learned that a U-boat had penetrated the defences and torpedoed the battleship *Royal Oak*. This had the effect of making us realise that we were not safe anywhere, hanging together, talking quietly of our setbacks, of the Germans' daring successes. Was our top brass decadent? Did they need a shake-up in men at the top? Hadn't Churchill warned us of the true ambitions behind the Nazi thugs?

Scapa, our stronghold, saw us entering and leaving. Sometimes the whole fleet would sweep across the North Sea in an attempt to stop a raider from escaping into the Atlantic or with a covering force for a convoy, maybe fifty miles away, sometimes actually steaming along with them, which was a wonderful sight. For it was our job to keep the raider at bay; the 'wolves' – U-boats – from picking off the stragglers, whom we could lead to the safe channels through the minefields.

Through those frozen winter months, in snowstorms, bitter winds, icy rain and freezing fog or the glorious starlit night with the Aurora Borealis spreading its streaks of silvery spray across the glittering starry sky, thankful for the warm woollen garments our womenfolk had knitted for us.

That Christmas was spent at sea, eating our Christmas dinner in relays of small numbers so as not to weaken our efficiency if a sudden raider was encountered; although midday, it was pitch dark outside. This weather suited the German surface ships in their escape into the Atlantic. This stalemate allowed each side to make preparation for perhaps a longer war than expected, the vast armies facing each other across the borders of Germany. France hadn't so far fired a shot, which seemed to us sailing the seas a bit phoney, but perhaps the delay favoured us, and the big guns and massive underground network of tunnels and storage depots of the Maginot line built by the French were giving the Germans a headache.

Now the Lords of the Admiralty had come to the understanding that a seaman gunner's course was far too long and complicated for an Able Seaman; the many guns and types that were now fitted on warships, the delicate instruments that control their accuracy, the big guns for surface-to-surface firing were too slow for the planes flying in at speeds of 300 miles an hour. So the 4" with their smaller shells, easier to manipulate with the changing range altering so rapidly that quicker fuse setting was required, which made the shell burst at various distances; then when this failed to stop the pilots, aiming their bombs at funnel heights, there was the multiple pom pom putting up a close barrage and numerous smaller weapons right down to the Lewis gun, even to our ·303 rifles. So swift action was needed to get the right sort of men, and the numbers to man the guns of the many warships, now being turned out of our busy shipyards.

Orders were issued from C.-in-C. to all ships that refresher courses be given to all gunnery ratings, with the choice of volunteering for the non-substantive rate of CR III (control) LR II (Layer), AA II (ack ack) or QR III (quarter). It didn't take me long to decide. For wasn't I a good shot? My catapult at home had given the family many a rabbit and pigeon pie. My keen eyesight had earned me the marksmanship badge I proudly wore. Being among the guns was much to my liking. Quickness of the eye and steadiness of the hand were part of my upbringing. I quickly informed the gunner's mate that I wished my name forwarded for a layer's course.

Then one bleak January day as we sailed into Scapa, when the freezing spray from the bows made me turn halfway sideways, drawing my head into my shoulders to escape from the full blast of the biting icy wind, how grateful I was when, task completed, I could rush down to the messdeck for a cup of hotters, looking anxiously round, knowing the mail should be aboard.

As we lingered around in the mess for our dinner to be laid out, someone bounded in, with a glow of excitement radiating from his beaming face, calling out 'It's Chatham for me, home' down the line, knocking the pints back as he grasped a messmate round the shoulders trying to make his partner do a kind of hornpipe; but he shouted 'Shut Up! I will have you run in for spreading false rumours.'

'Is it a rumour?', he chuckled.

Quick as a flash at these words up shot my head, but I kept calm under the impact of the sudden news. Was it genuine or a leg pull? There seemed a ring of truth in his voice, so I made for the notice board, where a small group had already gathered and was growing bigger every second. I struggled in my eagerness until able to read: 'The following Ratings will leave for gunnery course.' My eyes went down the list and there was my name. I turned away full of excitement, hastening back to my mess, saying 'It's true, and I'm one of them.'

Next morning a small number of us departed from the ship, bidding all our mess chums a cheerful farewell, catching the ferry to the mainland, travelling throughout the day and night, the wheels of the train singing the good news in my ears, as we passed through darkened stations of big towns. We changed trains at Glasgow, mixing with civilians for the first time since war was declared. The kind-hearted motherly ladies asking you to partake of a biscuit, the glow of warmth shown by all in their friendliness to the servicemen, made us aware of the concern and affection that now abounded in the land. Such was the transformation that had spread over the people in their common cause to fight this war.

It was late when I arrived at my old familiar barracks, and much to my surprise I was told I could go home on night leave as soon as the necessary formalities were completed. Even the hardened naval authorities were showing signs of humanity. So it was on a cold winter's night that I picked my way through the darkened streets to the outskirts of Gillingham. There to the utter surprise of my dear wife I announced my late arrival with a loud rat-a-tat-tat on the door, to hear the well-known footsteps pause to open and ask wondering who at a late hour came disturbing the neighbours. 'Locked the door against me, and I returned from the frozen North. Don't you recognise the warm hearted husband who sailed away?'

'And how is my little whale of a son?', I asked, after uncoiling from the reception and finding my voice again.

'A whale!' says my wife with delight. 'You wait until he's awake. He wriggles around like a tadpole.' Then she showed me my son for the second time, asleep, snug and cosy in his tiny cot.

'Is he always asleep?', I ask.

'You'll know if you stay long enough.'

'It's not an hour's visit like the last time. I'm back in barracks on a course for a short while, so I will have time to get all up to date.'

CHAPTER TWENTY-FIVE

It was great to be back a native again but alas, decisions had been taken. No longer could one plan for the future. It was thought better for the time being that the wife should carry on living with her people. We had given up our little bungalow which had been our dream home. It was best for safety, the bungalow being more in the centre of the town and nearer the dockyard, which, no doubt, would be a target area if bombers did come over. At present those at home sat waiting and wondering behind their blackout not knowing what to make of the delay. Those wishing could erect Anderson steel shelters as dugouts at the bottom of the garden. Those days, which were like living on a knife's edge, not knowing if air raids would come.

In the barracks many more recruits were being trained. The gunnery school was keeping up its efficiency but how long could this be maintained with the large number now being called up? No doubt the new system of training men to the similar types of guns would help, making it more professional in each group, for already the Navy was having some idea of what was required by the continuing skirmishes with the enemy.

My wife and I made use of the time given us. There were walks together, when we tucked our little son in his pram, and then his baths, where he would kick and splash and I would say 'Ah! He will make a true sailor!' What lay before us we dared not think or talk about.

One weekend I went to visit my people. I found the village very much alive; trade at the pub was improving with lively discussions on the war, usually starting after closing time, my father told me, and carrying on outside on the forecourt long after the doors had been locked. For many of the village men had been old soldiers; with a pint of bitter to loosen their tongues they would weigh up the situation, telling of their call-up, the first time they had left the village, their embarkation overseas and their experiences in the trenches.

It is always to the mothers that the fears of the war mean so much. 'We heard over the wireless your ship was damaged,' my mother told me. I tell her what happened as best I can and assure her we are well protected by thick armour and can look after ourselves. 'Take care,' she warns me, growing restless with fear and worry, which shows in her voice and looks. 'Don't worry, I am safe and well and shore-based. You must see your little grandson. We have named him David. You wait till we bring him down,' I tell her.

'I am pleased with the name. It runs in my side of the family. Your grandfather was David Free,' relating to me the family tree, forgetting about the war.

The course completed, I became a Layer Rating 3rd Class, and straight away was drafted to H.M.S. *Eskimo*, a destroyer of the Tribal class, one of the latest and larger destroyers. I joined her as advance party in the dockyard at Southampton. As soon as I stepped on board and saluted the quarter deck, I felt something of the strength and power as I walked amidships, over the top of the engine room. The destroyer's all-round appearance showed the true British characteristic of design that occasionally our skilled craftsmen bring forth as a world beater.

I dumped my hammock down on the mess deck and started to unpack, struck with the size of the messdeck for a destroyer—there was room to swing round. I selected my billet, to sling my hammock, a privilege being one of the first aboard, and somehow giving me an advantage over others.

The small number of us settled in to find that we were almost left alone, going ashore without the customary reporting and inspection. Even when called in the morning, we just turned over to carry on sleeping. When we did turn out it was the tea boat and a natter before deciding to have breakfast. But soon with an increase in numbers a stricter discipline was enforced.

Now the gunner's mate, who happened to be the duty petty officer of the day, thought it time to bring us back to the Navy standard of efficiency. After the usual pipe to call the hands had been sounded more than once round the messdeck and there was still no stir from sleeping sailors, the gunner's mate took upon himself to come and call, but what an awful voice he had. At last he left and we had won our victory, or so we thought, closing our eyes to sink back into oblivion. Some ten minutes later that rasping voice rang out again saying 'the last time'. Still no response, but soon after a different, quiet note of command came from the lips of the gunner's mate, which had some effect–for those who didn't obey, had their names scribbled down on a pad he carried, and they were told to turn out and fall in on the quarter deck, giving them ten minutes. So a bedraggled, sour-looking, heavy-eyed file of matelots shuffled aft to the quarter deck, myself among them, to the remarks of those returning off night leave, calling out that disobeying an order was mutiny and in wartime it carried the death penalty. Then: 'Off caps,' standing before the young officer of the day, facing the charge of disobeying an order, resulting in us all getting 'First Lieutenant's report'.

From now on the ship took on its full complement and the messdeck came alive. I was now a fo'c's'le man and shared the for'ard mess right under A gun. A leading hand of the mess took charge. We selected a caterer, the leading hand making out a list of cooks of the mess, two hands, whose duties lasted 24 hours from washing up after dinner to laying it out the next day, one from each watch.

It now became a very busy time for everyone. The deck was flooded, the big blocks that shored us up were knocked away with the stokers informing us that basin trials would take place while we seamen were provisioning, painting, cleaning with all special jobs being allocated out, the dockyard workers finishing off or being called in to put something right. At last we were getting shipshape and almost ready for sea trials, with all hands having to be put through their training in a very short time: what would have taken months in peacetime was now carried out in a week or two. Many times the gun's crew were piped to close up, learning their drill, practising to do things automatically while the specialist put to right any defects in our equipment.

Now the *Eskimo* ship's company of the lower deck had a sprinkling of the older hands, with a large number of younger ones to complete the crew. Nearly all had knocked around the world. Some of us already had a taste of action and now realised that our enemy was to be reckoned with. If only we could get to grips with his forces, to battle it out ship to ship. 'He will never do that,' spoke up Joe Blake, one of the old timers, as we sat discussing the war. 'Didn't we stay ready and waiting in Scapa during the last one, waiting for their fleet to come out and only once did that happen. That was when they had a fleet.'

'Lucky for us he didn't,' said Nobby White, 'our loss was the heavier.'

'Maybe,' replied Joe, 'but he scuttled back just in time, skipped away in the night, or our battle fleet would have knocked hell out of them in the morning. Jellicoe had them beat. That's why Von Scheer had to make a daring dash back to the safety of his ports. He knew he was beaten,' said Joe, with a knowing air.

'What's going to happen this time?', asked young Paddy.

'I'll tell you,' said Leading Seaman Baker. 'In his cool and calculating manner, he's either going to attack you from the air or torpedo you from under the water, with a hit-and-run raid every now and again to keep us tied down, which is what he is doing at the present, and if he can slip one of his pocket battleships through our blockade every so often, we won't be lying in our hammocks many nights. One of our jobs will be to scout round far out in the Atlantic owing to our longer range.'

'How long is this going to last?', asked Paddy.

'Last?' said Joe Blake, 'It hasn't started yet. Those little skirmishes some of us have encountered are just starfish stings to what we have got to face up to.'

'Oh, forget about it,' said Joe, 'I've got something will make you forget about the war. It's the old boys' game,' bending down and drawing out a cloth and proceeding to spread it out on the table. 'There you are,' he went on, 'the old game of crown and anchor. You play, I'll pay,' he cried out now that he had a small crowd around. 'It's no use saving your money. The old ship might go down. Life's a gamble. Here, try your luck,' offering the dice to Paddy to have a throw. 'There you are. Two crowns and an anchor. You would have caned me for five bob. Anybody for the old game?' He placed a pile of silver on the mess table remarking 'Pieces of eight. Any takers?', seeing the faces beginning to grin at his witticism, and now shaking so the disc rattled invitingly under the noses of his audience. 'You play, I'll pay,' he repeated.

We played for an hour, the war forgotten as eager eyes followed the dice, and Joe paid out or gathered in from his pile of silver and coppers and one or two crisp notes were added, before pipe down was called and Joe with a successful grin, folded up the cloth of gold saying 'enough for one night', and with loving care put the pile of silver into an old sock to stow away. The crisp notes he flattened out, then to his admiring shipmates sitting in the mess being highly amused at Joe's success, and intently watching, he fingered from round his neck a lanyard, loosened it and dangling down from the lanyard was a small leather pouch, which Joe proceeded to untie, with an air of knowing every action was being absorbed and curiously watched by his onlookers. He now took from the pouch, and unwrapped, a rolled skin cover, which contained a small round bundle of notes which Joe professionally straightened out, laying the two or three of the night's takings on top, to roll up and place in the transparent skin, tucking it in the pouch tenderly, easing it round his neck.

'I shall never get drowned,' said Joe to us. 'I bet none of you know or can guess what this is,' fingering the thin transparent skin in the pouch. 'While I have that round my neck I have no fear of drowning. If the ship went down, I should survive,' emphasised Joe, his eyes bright with the knowledge of having a secret.

'What is it then,' I asked, 'and what makes you know you will never get drowned?'

'You have a guess,' he declared, hanging on to his secret, enjoying the importance of holding our curiosity. 'It's a charm of good luck handed down from my grandmother and while I wear it, I shall never drown,' repeated Joe, tucking it away under his vest. 'That's where I keep it, round my neck. That's where it's going to stay,' he firmly replied. For Joe was a tough character as sailors go, and had done his bit in the boxing ring one suspected, when looking at his enlarged earlobes. Although not tall, his strength lay in his broad shoulders and we already looked on him as a tough handful if aroused, but gentle and full of good humour. One of those who never got taken out of their stride and got along with both officers and men, for Joe was the ship's sailmaker, one of the few left in this modern iron navy, and in his spare time, if one took along a roll of purser's blue serge, he would soon run you up a suit. There were other things Joe was teaching us, for we were the suckers who paid the tanners and the bobs and challenged the odds. There was a great deal of wisdom in Joe beneath that dark rugged appearance and good-humoured tolerance. He had now put away his takings, and seated himself on the table edge, so that he faced us, patting his chest where his mystic charm lay.

'Our object of curiosity—not one of you know what it is. Well,' he continued, 'it's called a caul. Now none of you know what a caul is,' said Joe, pausing as he eyed us over. 'Well, a caul is a protecting skin cover which babies are born in. If you get one undamaged, no harm will ever come to you. This one has been handed down from my grandmother, my father carried it and now I have it. It will never leave its place while I am at sea and while it remains round my neck, I shall never drown,' declared Joe with a firm declaration of religious faith in his conviction, and we sat quietly listening, fully convinced by Joe's statement. Then, breaking the silence I asked, 'Joe, will you let me wear it? Do you think I will get off my charge when I go before the First Lieutenant tomorrow morning?'

'It's never going to leave me,' he replied, 'but you can touch it and perhaps it may bring you luck.' I accepted the offer, fingering the smooth delicate cover that once held Joe's father, fully convinced in myself that this caul, as Joe stated, had some power of helping those in distress.

It was morning, the silence broken by 'Wakey Wakey, don't turn over–turn out,' the voice of the bosun's mate and the shrill call from his pipe stirred the sleeping matelots. Most of them roused, stirred in their hammocks then carried on snoring. One or two turned out, someone's dangling legs upsetting the mess utensils, the jangling noise bringing out a muttered oath from a sluggish sleeper. I woke up, lay thinking for a minute, then not saying a word I swung myself out, lashing and putting away my hammock rails.

'Put your head in this,' said a messmate dangling a rope with a noose in it.

'What for?' I asked.

'It will save all the trouble of hanging you at sea. If you let me hang you now, you'll get buried on shore. You know the penalty of mutiny in war. That's the price you'll pay this morning. "Stripes" in the other mess has been practising the hangman's knot and we might as well have a trial run to see if it works. You'll only be hung at sea and ditched overboard as food for the fish. Not even a prayer will be said over you out there.'

'What–do you think I have just joined?', I retorted, 'Not guilty will be the verdict, just you wait and see,' I remarked, as I went about my duties, not at all pleased at the sarcasm being shown my some of the messmates. By ten o'clock, I had my defence all planned and sewn up. Inside I had already carefully gone over my story, had ready my answers if there were any questions and my excuses, thinking it was fairly foolproof, taking into account we were not all known to each other as none of us had been on board more than a week or two.

There we all were, lined up in our number ones, the table placed on the quarter deck, with the First Lieutenants standing over, with the other one or two ship's officers present. Not so awe-inspiring as on a cruiser but all the same, a respect to justice, as the coxswain snaps out his command, bringing us all to attention, saluting the First Lieutenant, then proceeding with the court. One by one we were called up to face the charge. 'Off cap,' and I could hear 'Three days number eleven' being given as punishment to each individual. 'On cap, double march,' as each turned, doubling away to lament his grievance. My turn came well down the list. Could I remember my alibi? Nearly all seemed to be taking their punishment without a word. Now my turn. I double up and salute. 'Off cap.' The charge read out: 'Refusing to obey an order. Not turning out of his hammock when ordered,' read the coxswain. The writer on my right, with pen in hand ready to tick me off and the gunner's mate ready to give his version.

'Well, Sir,' I spoke up, 'I had turned out early to scrub the wardroom out, which I had done, went back to lash my hammock up and along came the duty Petty Officer, Sir, and took my name.'

'Why hadn't you stowed away your hammock when you turned out?' questioned the First Lieutenant. 'It was not possible, Sir, in the position my hammock is slung. I disturb so many others. I have to go back later. Your orders are, Sir, the wardroom to be scrubbed out before breakfast.' I stopped. I must not say too much but just enough. Had I convinced them? I saw the writer stop with his pen poised while the First Lieutenant paused to consider. Would he call on the gunner's mate? Then he spoke. 'See this rating gets a mess change to the watchkeeper's mess. Case dismissed. 'On cap,' ordered the coxswain. I saw the gunner's mate's long face of woe standing right by my side and felt a bit sorry for him at my untruth as I doubled away smartly in triumph, but careful not to display my good luck as I was the only one who got off. Bad things about my reputation could go round but the mess change more or less covered that but I whispered to Joe that evening that I had every faith in his lucky charm. 'It's magic works, Joe,' I told him. For I was the only one to touch that charm and the only one to get off.

'There's no doubt about that. Come what may, I shall never get drowned,' confirmed Joe.

CHAPTER TWENTY-SIX

Now we were guarding convoys a long way out in the Atlantic, meeting up with the incoming ones and escorting them right up to the harbour entrance of Liverpool, fussing and rushing around as if impatient at the slow-going merchantmen, laden down, moving so slowly and such a large target.

But Scapa was our rightful base, being attached to the main fleet, whose duty it was to guard England against the powerful German warships. Now that we had got our sea legs riding the Atlantic rollers, it was the North Sea we had to face up to; if you can face up to the winter's storms on a destroyer in the North Sea, you can face up to anything in the world. Joe told us this when we had a break to spend a day in harbour, bailing out the messdeck, collecting up the utensils, scrubbing and drying out or replacing the damaged crockery, but the harder and more dangerous your life, the more you enjoy the smallest comforts.

One day there was an empty place at the dinner table. Alarmed messmates reported their mess chum missing; there was a pipe around the ship for Able Seaman — , a search in the overcrowded ship, a shipmate's last report of seeing him on the upper deck just before a big wave struck. After a final visit from the First Lieutenant and coxswain, the ship's log recorded 'Able Seaman — missing at sea, believed drowned'. Deep sorrow hung around us at the fatal mishap, as we saw how the relentless sea in its fury could snatch a victim from your side in a crowded ship, unsighted and unseen.

We didn't always know what was going on or guess what lay behind our operations, although the observation of every change of course led to rumours and each rating had his ears and eyes glued to catch a look at the charts or half a glance at the latest signal, on refuelling and leaving harbour.

One day all hands that could be spared were ordered to muster on the messdeck, to be told by the captain that we were all sharing in the secret of this operation. Our secret service had obtained information that the Germans were going to seize Narvik, the most important Northern port of Norway in the Arctic Circle, where a large amount of iron ore was shipped. Our orders were to seize the port first, before the Germans arrived. We were to sail in alongside, rounding up and capturing the captain, officers and crew of all the merchantmen that lay in the harbour, before they had a chance to give an order to get under way or scuttle their ships. Surprise would be the key to the success of the operation. Volunteers were required for boarding parties. They were to be armed with cutlasses and revolvers. It would be a dawn attack with other destroyers. As soon as we got close enough the armed raiding parties would leap aboard, certain ones to rush along to the bridge, thus stopping any ship from getting under way, while others raced along to cabins, rounding up the merchant seamen before they had a chance to realise what was happening. Nobody was to be shot unless they showed resistance. Bloodshed was to be avoided if possible. Anyone hesitating, a prod or two with the cutlass would be effective enough to hustle them along and you know where the tenderest part is! The First Lieutenant was to lead the boarding party, who were to make themselves gruesome with dabs of paint and soot, wearing any old rags. Identification would be by a coloured armband given just before boarding. Prisoners were to be taken along to the ship's side, where hands would be waiting to take them over and lock them in one of the cabins. Riflemen would man all vantage points in case of unforeseen trouble. A skeleton gun's crew would silence any shore battery. It would be a dawn attack, and there were two or three days for pistol practice and cutlass drill. 'Give your names to the coxswain, those who wish to volunteer, and the First Lieutenant will straight away get down to organising each man's part.'

Immediately there was a rush to volunteer. I was amazed to see the enthusiasm and eagerness shown by all. In no time at all the lads picked for the armed raiding party were rehearsing. The chef was ordered to stoke up the galley fire to get soot to rub on arms and faces. Bending over to adjust his fire nozzle, an armed assailant couldn't resist the temptation offered to try out the point of his cutlass on chef's tenderest part! A yelp – with a spiralling spring of sudden velocity that shot him up, hitting his head on the boiling saucepan handle, causing a hiss of enveloping steam, with chef emerging brandishing a red-hot iron. The startled tormentor fell backwards, losing his cutlass and fumbling to draw his pistol as the chef dwelt resounding blows, causing burning sparks to fly in all directions. The First Lieutenant, unaware of the real cause, shouted 'Stop making it too realistic.'

He thanked the chef for demonstrating his prowess, but warned him that his burning desire to use a red-hot iron could result in casualties before the real offensive began. Off to such a rousing start, there was no holding our raiding party. With skull and crossbones, painted red over black faces, artificial pigtails made out of spun yarn tallowed down to make it look as though the ghosts of the cut-throat boarding parties of the old sailing men of war had returned. The clash of steel and the crack of the pistol shot could be heard as they rehearsed in the two following days, but then a disappointment that would have dampened the spirit of Nelson, much more the ship's company of H.M.S. *Eskimo* – a signal 'calling it all off'.

On land the war was at a standstill, only at sea was constant activity kept up, and that seemed to be by destroyers of the Royal Navy. Then suddenly it took a dramatic turn, in the early days of April. The Germans invaded Norway. The mountains and fiords we had seen at a distance now became our hunting ground as an escort to the cruiser *Belfast*, saw us cautiously entering the northern fiord round Narvik, the snow-capped mountain peaks looking like sentinels silently watching as we slowly nosed our way through the still waters of the mountain-locked fiords, looking for hidden enemy ships but so far not finding any. Although the winter months had been the roughest for years, now the water was still with a millpond smoothness, the sky blue and clear, the sun's brilliance reflected from snow-white slopes. It was so peaceful and quiet, with the wondrous beauty of the towering peaks. But still waters can hold hidden dangers. Cautiously we steamed through the jagged rock-infested coast, sending spreading wavelets rippling across the smooth waters. Looking across at our cruiser, I suddenly shouted out 'Look, they have gone mad!' – for every man on the upper deck was rushing around seizing everything movable and throwing it overboard. 'What the hell are they up to?', I asked as 'A' gun crew stood amazed, watching the frantic actions of the sailors on our parent ship. Then from the megaphone the captain's voice gave the order 'Stand by to tow.

Belfast is stuck fast on a rock and doing what she can to lighten herself.'

Myself and others on the upper deck rushed along, unwinding and laying out the wire hoses and all equipment needed to tow. The captain manoeuvred his ship close enough to pass over a line. Quickly this evolution was put into operation, a credit to our seamanship and peacetime training, and the badly crippled *Belfast* was hauled off and towed through the narrow entrance of the fiord to some sheltered cove surrounded by high mountains, the searcher now becoming the hidden ship. We heard afterwards that the pilot who had been picked up locally at the request of the *Belfast* and who was responsible for navigation, had been instantly shot dead by the *Belfast*'s captain. This I believed to be untrue.

Joining up with another cruiser and destroyer, we continued our searching operations unaware of all the activity carried on by other units of the fleet, along the whole Norway coast, on the late afternoon of the twelfth of April. This cruiser escorted by the *Eskimo* and another destroyer had orders to sail down the inland stretch of water to Narvik. How quiet and still sailing between the lofty mountains, the guns loaded and pointed ready to fire at a touch, every man alert for instant action, not a sound from the three ships; above the sky so clear; not a stir from the snow-laden branches of the pine trees. Men standing by their guns like statues, waiting to act, their life depending on swift retaliation. The magnificent cruiser with her two guarding destroyers, boldly sailing down the landlocked stream, where hidden guns could be concealed in clefts of rock, or hidden in the snow-covered fir trees, or the waters mined. What was going to happen? The only movement that I noticed was the sweeping search through binoculars by standing officers on the bridge of the advancing ships. The still serenity of the northern peacefulness. Then momentarily a flash of flame; a spiral of water shot up just off our side. Not a flicker from the mountainside. Not a stir from the frozen land as we waited. Even our hearts stopped in the nervous tension. Then mysteriously, and without a sound of what it might be or where it came from, another spiralled up, just over our bows. Breathless long moments before flags fluttered from the cruiser's yardarm. A signal, then I became aware the *Eskimo* was turning, the churned-up water as the three ships heeled over, helms hard over as we turned in the narrow channel, steaming out as stealthily and silently as we steamed in. 'An electrically controlled minefield,' a member of our crew suggested as we sailed back, our normal feeling returning after the tension of the unknown advancing danger, returning to the sea entrance, biding our time underneath the mountains of the coast until the following morning of Saturday 13 April 1940, the sea becoming unsettled as we hung around. I awoke from my doze in our huddled quarters of 'A' gun's shield, the visibility reasonably clear so we were allowed to nip below one at a time, taking the opportunity to dip my head under the bathroom tap, with ear alert to race back at the first note of alarm. That morning, as I walked back feeling the *Eskimo* gathering speed, I had just jumped into my trainer's seat as the alarm sounded, following the red pointers on my training dial, wondering in the rapid seconds what it was. Then the loud report of depth charges, shaking the speeding ship. 'What's happening,' I asked as we slew round, steadying up to make another attack. A U-boat had surfaced, flashing a signal in morse code. We had just smothered the spot with depth charges, after she did a crash dive, I was told. I felt annoyed to have taken part in this incident, yet not seeing anything of our enemy. No trace of oil appeared on the surface to let us know the result of our attack.

The First Lieutenant addressed us: 'Well, boys. We are waiting for reinforcements to come up, which shouldn't be long now, a number of destroyers with the battleship *Warspite*. Then we are going to sail into Narvik and give battle to a flotilla of German destroyers in there. The *Warspite* will be following us up to silence any shore batteries that may be defending the harbour. May God help us and good luck', were his hurried, departing words. I looked at the ring of faces of the men, told they were about to go into battle. No fear of death was showing, no sign of cowardice. Some appeared to welcome the action. I had a sinking feeling, and couldn't help admiring others, those who would be working in the deep holds of the magazine, handing out the shells through the hatches along the messdeck to the upper deck, keeping the supply going to the firing guns. Our skill that had been taught us by the gunnery school on models and weapons, was about to be put to its supreme test of life and death. The navy tradition of steadfastness was to be upheld. My God – I'm going into battle; it's real. I was thinking that for all my life I had wondered what men had felt like lined up and waiting to kill each other, the thoughts running through my mind in the minute or two before 'action stations – clear away for battle'. The pipe we had often called in fun on the messdeck was now reality, and when I stepped on to the open deck, there were the reinforcements approaching, the White Ensign proudly steaming from the mainmast. A rating in charge of 'A' gun's magazine rushed up to take a last look round at the open sea, seeing the approaching ships race back, shouting excitedly that reinforcements had arrived.

Every man was at his station, myself giving a final blow to my lifebelt, tucking in my anti-flash helmet round my neck, tightening the arm straps of my gloves, consciously aware of my nervousness as I mixed with 'A' gun crew, our leading hand, captain of our gun, giving us comforting and encouraging advice to keep our gun firing, an all-round check; make sure the upper deck ammunition was handy and ready, myself looking through my training binoculars making sure they were clear. Now the destroyer flotilla was forming up in battle formation, according to seniority of rank. We were fourth in line, about to sail up that same stream that only a few hours ago we had turned back from when mysterious splashes had shot around us. There was the fear I had of sailing into a minefield. Looking through the opening in the gun shield, seeing the stern of the destroyer ahead churning up the water, her quarterdeck gun's crew moving around. Leaving my gunshield and looking astern, the line of the advancing warships, the twin barrel of the foremost guns of the following destroyer pointing outwards, the bridge dotted with standing figures, watching ahead. Down the line of the advancing destroyers to the squat mighty *Warspite*, bristling with guns bringing up the rear, keeping station as they had done so many times before in peacetime manoeuvres, sailing up that smooth silent channel engulfed by the mountains, aware of the uneasy stillness, steaming gracefully in line ahead, with the White Ensign spread from the masthead. The Germans has destroyers with 5.5" guns, slightly larger than our 4.7" ones, but we had a battleship to back us up. Would she be able to manoeuvre in a small space? What secret information had been passed on to the Admiralty for them to risk a battleship upstream? What if the German army had brought their guns and lay hidden behind large boulders of the rugged high banks of this waterway? All these thoughts were passing through my head as I waited around with other members of the gun crew, saying very little, perhaps like me thinking and having a nervous sinking in the stomach. Now the order: no one to leave his position. I settled myself into my trainer's seat, gripping hold of the training wheel, watching for the movement of the red pointers, aware that my white pointers had to be steady and in line the moment the guns were fired, each member of our gun standing alert, ready to commence, with shells grasped to reload. The tense minutes, the sinking nervousness, then slowly my training pointers moved, only to go back to zero. Information came through that the Germans has left harbour, coming out to meet us, waiting so quiet with my eyes fixed on those pointers, watching for the slightest movement. 'Enemy in sight.' A breathless moment, then the boom of a gun. Our leading destroyer had opened fire, commencing the battle. The director pointers moved round. I followed, keeping close behind with my white ding-dong of the fire bell. I was steady and in line as the blast and shudder of the broadside sent the pointers leaping. I heard the clatter of the empty cylinder falling on the deck, the hurried reloading, the snap of the closing interceptor. 'Ready, ding dong.' I had the leaping pointers steady as the broadside roared. I knew nothing else, my full concentration on my training dial, the ding-dong recoil of the breech with the flash and roar were automatic. I gave all my willpower to my job, with a prayer to my maker. Before, I prayed without much meaning, but my God, I was asking in earnest now, for the screeching of the enemy shells tearing through the air was reaching my ears between the roar of the broadsides. Did I dare take my eye from my training dial? I took a momentary glance. A shell splashed close by. I concentrated on my training, the cracking gun vibration and the din continued. A sweep round by the director, the firing eased. I took the opportunity to look up through the gun shield and could see a German destroyer bottoms up and smoking, but then the ding dong and we were firing fiercer than before, the flashes from the barrels were burning my face. I noticed that the paint was peeling off the red-hot barrels, screeching shells were exploding and falling around. The shout of the Captain of the gun for more shells, a boy whose voice I recognised yelling 'I can't go on'. 'You must,' I heard our gun captain shout, then the crack of an exploding shell, the curtain of flame and roar of our gun shaking me in my seat. Then from the communicating number came the message 'Very good laying and training from A and B guns', as the firing slowed down, then stopped, the order being a lull in the battle. I sat back to look through the open trainer port of the gun shield, seeing another destroyer driven on the rocks, lying on her side, seeing figures scrambling off. The Captain of the gun turned to me, asking if I was all right. I answered 'Yes'. 'Take one of these,' offering me a cigarette and giving me a light in quick succession. Then I took a look round, but could not see very much through the smoke. It seemed that the channel must have opened out into an inland lake where the battle was taking place. A cloud of smoke was lying across the water, which we began to follow, coming across a smoking canister every few minutes. They appeared to lead to another small inland creek as we nosed our way hot on the trail of the fleeing enemy. 'Watch out,' called my leading hand as the pointers on my training dial moved round. 'All right', I answered, 'I am steadily following', for the smoke was thinning. I could see a bend in the river which we had to negotiate. Slowly the bows swung round, as I realised we were the foremost ship and being on 'A' gun, the foremost men in battle and clearly as we rounded that bend, two large German destroyers, no more than a thousand yards, lay broadside, two waiting at another bend. I saw the flash of their guns as they fired, point blank at us, as our bow slowly came round, a group of shells from their broadsides dropping right in front of us. The next one, I knew, would hit us. Why, why don't we open fire, was racing through my brain. For God's sake open up. Seconds that seemed like years, the imprint on my mind of two long German destroyers stretched across the next bend of the creek, the flash of their guns, a group of shells dropping right in front of our bows, a frantic prayer. Hurry up and fire! I knew I was going to be hit. Then the flash and thunder roared in my ears. The firing pompoms, the cracking rattle of the point fives were all distinguishable, telling me we had opened up with every firing gun we had. The fierce onslaught, noise, shudder, flame and smoke. The sudden vibration caused by the propeller, all registering in the deafening roar. Why, why is the ship going astern my mind was asking? 'Where am I, where am I, where am I? hearing my voice repeating, faintly as if it was a long way off. I saw mountains reaching ever so high, trees that stretched further up the sloping sides in some places than others. 'Where am I?' Lost and bewildered, my mind struggling to overcome its dazed wondering condition. Slowly a warm flood of realisation was gradually spreading through my brain. I was remembering. I heard my voice saying, sounding closer and clearer, 'I know, we were in battle'. I burst forth, perhaps a bit excited with realisation of my memory becoming fully restored, aware I was now standing on the deck amidships, a shipmate on either side, holding me up. 'What happened', I asked.

'It's all right, it's all right. It's all over,' they replied. 'We are waiting to be towed away. The ship's been damaged. The bow almost cut away by a torpedo.'

'Oh,' I yelled out in pain. 'My leg's hurting.'

'Where?'

'My thigh, right here,' putting my hand where I was feeling as if a red-hot poker was burning my flesh. Immediately one of them took out a knife, ripping off my trouser leg. I became aware that my clothes were wet and a bandage was bound round my head with a label attached, but all there was on the fleshy part of my thigh was a graze a few inches long, but how it hurt. 'Take him along to the after lobby and get a bandage on that,' ordered the Doc, taking a glance as he rushed by.

A dreadful sight met my eye as I was led along to the after lobby to await a dressing. There, lying on stretches and blankets were many wounded and nearly all of 'A' gun's crew, hardly recognisable for their bandages. Our leading hand had a big gash in his head; so white and still he lay that I settled myself down, bewildered and quiet among the bodies. Some stirred and moaned, some never made a sound. Had they passed over? Soon after the Doc came round, examined me. 'Where was your action station?' he asked 'On 'A' gun,' I replied. 'You must have been born under a lucky star', was his reply. 'Now I'll bind your leg up and your shipmate here will look after you. Go down below and find an empty officer's cabin. Take those wet clothes off. You'll find some dry ones. Help yourself, and turn in; sleep if you can.' Down below I went, found an empty cabin which happened to be the gunnery officer's, stripped off my wet clothes, opening a drawer, finding and putting on the gunnery officer's silken set of underwear and for the first time in my life, turned in on an officer's bunk. There I lay, bewildered and partly dazed, wondering, feeling so lonely, then suddenly the alarm bell rang. I heard the scamper of many feet, the cry of enemy aircraft approaching. Then silence. Tense, frightened minutes of terror, lying wounded, not having anything to do, just waiting to hear an explosion. All so silent above as long frightening minutes dragged on, ending with the most welcome words, 'Aircraft passed over'. I said a prayer of thanks. How long would I be left, for it was dark and lonely lying and thinking. How many of my gun's crew had been killed? The captain of the gun lay so silent with that large gash in his head. His position had been next to me. After a long, long time, a shipmate came down to tell me the *Warspite* was sending a motor boat to take off the wounded. Lying hopefully listening in the long dark hours. At last the sound of an engine, the slowing down, hearing a boat draw up alongside, the hurried, quiet footsteps above, as a voice almost in whispers saying, 'Lower, steady, careful right', and so on and I guessed they were moving those who were badly hurt, revving up as the motor boat sped away. The anxious wait for its return, then a shipmate came down. I wrapped a blanket round me, and was hurriedly and silently assisted into the waiting motor boat, a boatful and we were away through the darkness, relying and trusting so much on others to find the way, thankful at last as we drew alongside and clambered aboard, at last feeling safe on the mighty battleship *Warspite*. There we were made as comfortable as possible, one of their large messdecks being cleared and some seventy wounded men from the destroyers being laid on stretchers and blankets. Then I suppose I must have fallen asleep for the next I remember was waking up, the *Warspite* under way, seeing the many badly burned bodies lying around, pus oozing from eyes and lips, but it was surprising after a day or two, bandages removed, life and sight returning to some who I thought would never recover. We even had a German among the wounded, being treated and given cigarettes the same as us.

Some of my shipmates were walking about attending to others and I was given a fuller account of the battle. The captain of my gun died the next day. Another member had both legs blown off. Both their positions on the gun had been on my side. The leading hand's last act to me had been to lean over and give me a cigarette. He was a married man with two children and lived in Gillingham. I must have been, as the Doc said, born under a lucky star to get off so lightly.

In the heat of the battle, when I had felt the vibration of the propellor, causing the ship to go astern I had wondered why the captain was giving the order 'Full speed astern' to avoid a torpedo. Unable to get clear, the torpedo had struck us in the for'ard magazine, just astern and under 'A' gun. Fortunately the magazine was empty, for'ard ammunition all having been used in the battle, but the explosion had ripped through the ship, leaving the bows and 'A' gun hanging in the water.

'How long do you think the battle lasted?', I was asked.

'Twenty minutes?', I hazarded.

'Five hours and twenty minutes' was the surprising answer.

'I suppose it must have been if we had used all those shells up,' I replied thoughtfully, remembering those red hot barrels with the paint peeling off. A shipmate told me afterwards he had pulled me unconscious from the water, with pieces of shrapnel embedded in my forehead. The doctor came along with his tweezers, pulling them out, and he still has a piece at home, keeping it as a souvenir. A large number had been killed and injured, the *Eskimo* later being towed into a small fiord along with other ships.

There Joe told me he had the daily task of recovering the dead, sewing them up in hammocks. Later they were attacked almost daily by German bombers, the captain, parson and gunnery officer making up the gun crew and manning the pompom, firing to keep them off, the tall mountains of the fiord which they had named 'Crocks' fiord' making it difficult for the attacking planes to hit their target. Thus they survived for several weeks, facing the almost daily bombing and committing the recovered dead shipmates to a watery grave, weighted down by the few unfired shells, until they were towed back across the North Sea, stern first to Barrow-in-Furness. There the remaining bodies were recovered, the *Eskimo* having an intensive refit, Joe's mystic charm living up to the faith that he believed in, for Joe never got hurt or drowned.

Back on *Warspite*, with some of the injured recovering from their wounds, beginning to sit up or walk around, a few days and nights and the hopes of safe hospital treatment and seeing our loved ones soared, when at 1800 hours it was broadcast 'Expected arrival at Rosyth in the forenoon; all wounded will be taken off.' A cheer went up from the stretcher cases, their eyes shining at the good news. Letters were stamped and collected for censoring so as to catch the earliest mail off, because the Nazi broadcastng propaganda lost no time in giving out the names of ships which they claimed to have sunk or damaged, a certain percentage of it being accurate, thus keeping a number of British people tuned in to their news items. That night I rested, trying to sleep as the battleship rolled in the swell of the sea, with the thoughts of loved ones and home running through my mind, greeting my wife, telling her the good news that I had been spared when those by my side had been taken away. Of course I couldn't sleep! Lying and waiting for the long night hours to tick by. Then the buzzing of the loudspeaker, about to announce, my ears alert. 'Do you hear there.' My breath intake stops. Waiting... 'The *Warspite* is returning to the Narvik area.' My soaring spirits sank, the dismay shattering my thoughtful hopes. For the large battleship was altering course, returning to the Arctic region. I lay back disconsolate. Someone nearby whispered 'Did you hear that?' I nodded too choked to speak, for there were some amongst us who badly needed careful nursing and would have to be told in the morning that the landing had been put off for a few days.

Ten days after the battle, finding ourselves back in the snowstorms, viewing the white-capped mountains off the Norweigian coast, we anchored in a lonely bay, taking in oil. Also anchored in this bay was a large liner having a thousand troops on board. 'When these are landed', we were informed, 'those who are well enough are going to be transferred to the liner for passage home.'

So next morning with a slight sea running, we saw boat loads of soldiers being ferried ashore by the lifeboats of the liner. Later on, we that could help ourselves were picked up from the *Warspite* and taken to the liner, climbing the many steps of their gangway, and what a welcome we got. White-coated stewards conducted us along a labyrinth of carpeted corridors to cabins of single, double and family size, bathrooms attached with hot and cold, fresh or sea water. But that was not all. To our amazement – those who hadn't been aboard a modern large liner – the name we learned was the *Franconia* of 25,000 tons.

'When you have settled in and taken a bath, I will call and show you down to the dining hall for lunch, which is served between the times stated', giving us a card with the times of lunching and dining. 'I am your steward,' the dignified, smart white-coated sailor explained to us, 'and in charge of cabins of certain numbers. Any requirements, just ring for me. A bottle of brandy, rum, whisky or gin is very cheap, you do not have to pay tax on it. All the sailors will be off tonight and I promise you fellows we can have a grand time.'

I couldn't believe it, or take it in. To find myself suddenly transferred from a stretcher on the heaving deck of a warship, into a first-class cabin and the service of a luxury liner, being treated as if I was a titled paying passenger. For when my steward left me, having turned the bath water on, laid out the towel and soap, saying he would return in 30 minutes to conduct me to my dining table, I had to pinch myself to see if I was in this world, for I had a feeling that perhaps I had been killed in the battle and was really in heaven.

It was after my bath and wanting to shave and comb my hair that realisation hit me, for I hadn't a razor or change of clothes, only an old pair of trousers and jersey that some kid sailor had given up on the *Warspite*. All my personal effects had gone, lost for ever. The photo of my wife, which had always greeted me when I opened my ditty-box lid, personal letters which I treasured, the piece of gold and silver ore from the mountains of the Andes. Yes, I was stll on this earth, the bandage round my head wasn't a halo. Someone had been given a razor and was passing it round. A hot bath and shave, with the gunnery officer's silk pants and vest turned inside out, I felt dressed for the occasion, giving the steward full credit. His manner of conduct was the highest, with that class of dignity that only trained servants of cultured manners could carry out, unperturbed by his shabbily dressed and flustered guest, who followed him along corridors of the cabins to a large panelled interior dining room, all set out with tables and chairs, the tables being laid with serviettes, gleaming cutlery and sundries, the soft tread of my feet on the carpeted deck, the quiet-speaking and faultless waiters bearing trays containing silver dishes of hot courses, being served on our plates, made it hard to believe we were anchored in a bay of an enemy-occupied country. It was as if we were sitting in a grand hotel in the centre of fashionable London.

After we had bathed, lunched, slept and dined, the same tourists on a first-class cruise, our steward, true to his word, called us together in the family cabin, opened bottles of whisky and gin, filling glasses and freely offering round, clinking glasses and settling back. We praised our steward for the courtesy and treatment given, who in turn, apologising that our first-class entertainment stopped at dinner, related to us stories of wonderful cruises with famous people. 'For it's after dining, later in the evening,' he carried on saying, 'when the fun starts.' The steward carried on in his easy manner and entertaining voice, giving us names of famous stars and titled people that booked for cruises and had been given cabins under his charge. It was well past midnight when we broke up our small party, our plight and condition practically forgotten while listening to the steward's fascinating talk. I think all of us had our minds filled with a sunny cruise and glittering night entertainment, the ship now empty of the army, there would be more parties, he assured us, saying how sorry he was not to be able to provide the ballroom–dancing ladies. Then we retired to our cabins like officers, our minds partly taken away from our recent ordeal.

But alas, the unfortunate soldiers who had been ferried ashore, finding the land covered in six feet of snow, nowhere to sleep, knowing not where they had to go and having no transport, were fortunate in having a brilliant officer in charge, who decided straight away that this was no place to start an operation, and since the liner was there, he got them back on board as quickly as possible, and our luxury liner decks and corridors rumbled to the returning tramp of the thousand troops, cutting short our private parties and the individual attention that had been lavishly given us, by the merchant seamen of our sister service.

All that morning the re-embarking continued and very soon after the last boat had been hoisted and the gangway taken in, it was up hook and away, well out to sea and setting course for U.K., with the Arctic night enveloping us in its darkness. The liner made fast progress with her precious cargo of humanity and I lay in my cabin wondering what chance of survival there would be for a thousand-odd souls if torpedoed. But it was the second night out when a loud explosion woke me up. I felt the liner turning, gathering speed to race away on a different course. It couldn't have been hit, I knew. I never could find out what it was although many had been awoken by the shattering noise. A few days steaming, leaving Norway behind and getting nearer to home, our thoughts soaring again as we walked up and down the deck, enjoying our easy life, but one night I had a very troubled sleep. In a dream I saw my mother trying to tell me something. Then down came a cloud of mist about her, lifting her to slowly fade away, leaving me unable to understand her message, but all the next morning the vision worried me and I couldn't shake it off.

Another night; only a number of hours and we would be sailing up the Clyde, 'landing at Greenock' was the information passed around. I would soon be home or in communication by letter, my wife and parents would be wondering and worrying for it must be over a month since last they heard. Soon I would put their minds at rest.

Next morning, sailing up the Clyde, dropping anchor in midstream and being ferried across to Greenock. Once again standing on solid earth, the thankful feeling that I had been saved, to return to those who waited. On the jetty four or five ambulances were drawn up and in there we were packed to be driven off. But more surprises were in store, for it wasn't a hospital, navy or military establishment we stopped at but continued through the town into the countryside, turning into a sweeping drive with borders clustered with shrubs and trees, pulling up in front of an old mansion, where nurses in starched blue frocks came out to meet us. This was Ardgowan House, set in broad acres of parkland in the village of Inverskip, on the South bank of the Clyde, some seven miles from Greenock.

'If you follow me', said the cultured voice of my young nurse, 'I will show you the rest room at the end of this hall. There are golf clubs if you wish to play,' pointing to a golf bag in the corner. 'There is a fifteen-hole course on the West side of the park.'

'I often played in my village at home,' I replied, which was true, thinking of my boyhood days when we found the golf balls lost by the college gents and an old club they had thrown away, practising our swing, shading our eyes, shouting 'Fore' when we hit the ball.

'You can rest here,' she continued. 'Read a book or write your letters and now we will supply you with some fresh clothes.' I followed the nurse to the room where a large heap of clothes was being turned over for us to pick and choose, helping myself, but failing to find a suitable jacket to fit my slim shoulders and having to take a jersey, which drowned me.

Doctors arrived to examine us. I felt all right, although I believe I must have been badly shaken and there was a rash coming out on my foot, which the doctor seemed to think was the shock coming out, which puzzled me. How could shock come out in a rash on my foot? My head wound was healing and bandages removed. I thought I was fit and well but the doctor seemed to think I needed further rest and treatment.

What better place than a large mansion on the banks of the Clyde to recuperate? Hills to the North, wooded banks, boats sailing up and down the river, the old trees. Inverskip, the village, with its stone-gabled cottages. I felt at home in this lovely country and the scent of the heather was a tonic.

In this stylish setting I tasted the grandeur of a duke's life. After being served an excellent breakfast I settled myself in a comfortable armchair in the rest room, reading the morning paper or an illustrated magazine. Then a morning stroll through the glens and over the hills, returning to this ancestral home, where one of the ladies, ever ready to ask after your welfare, would tell you to rest in a chair while she fetched a cooling drink. Leading a life of a gentleman and being treated as such was beyond my wildest dreams.

My thoughts of home were never far away, wishing my wife was here to share my convalescence. She would know by now where I was, safe and recovering in my gentlemanly life. One day, returning from a walk through the park, I lay resting by the window when one of the ladies walked noiselessly over the soft carpet, stood over me and softly coughed. Then she spoke quietly saying 'Excuse me,' then asked 'Are you Able Seaman Sanderson?'

'Yes,' I answered.

'Will you please come with me? Matron wishes to have a word with you.'

I began to feel a little scared of being taken before the head one in authority. I hadn't broken any regulation and my nature wouldn't allow me to do any wrong where only noble respect and treatment had been given to me, so it was with a light step and wondering look I stood before and searched the face of the matron to see if there was that look of annoyed severity on her face, but that kind soul offered me a seat, then turned to close the door of the room behind my retiring escort, walking back to look me in the face.

'You are Able Seaman Sanderson?'

'Yes,' I replied.

'Where do you live?'

'In Gillingham, Kent,' I answered.

'Are you married?'

'Yes.'

'Have you parents who live in Cambridgeshire?'

'Yes,' I replied, 'my father and mother live at Great Wilbraham in Cambridgeshire,' getting rather uneasy at these personal questions.

'I have a telegram here, I am sorry to say, with some bad news.'

I remained motionless as if something was draining from my head.

'Your mother has passed away.'

I sank to my knees. Some minutes later that kind matron, her hand on my shoulder, was saying 'I will see the doctor for compassionate leave; you can be driven to Glasgow to catch the first train to London early in the morning.'

'Was your mother very old?'

'No, no, only sixty-two,' I chokingly replied.

'Had she been ill?'

'No, my mother was all right, she shouldn't have been taken away,' I burst out, anger overcoming my sorrow, resentment filling my heart, as I began to realise I would never see my mother again. 'She had many years in front of her yet. Why should she have been taken?'

'There, there, sit back for a while. I will leave you for a few minutes and come back with a drink.' She returned some five minutes afterwards to find me quietly settled, my resentment bottled up and thankful that one so understanding as matron could calm my sorrowing spirit.

Next morning I took my seat in the train sitting in the corner of the compartment, content to be left in solitude, wearing my double-size jersey, showing my vivid scars of battle. Many looked at me curiously and I wondered if they thought I was a planted fifth columnist. One kind lady gave me some biscuits. We reached London, and I found my way to Liverpool Street Station, having to wait for a train to Cambridge.

I went into a café set aside for servicemen, where tea and cakes were cheaper. The waitress in the cafeteria looked at me and then asked 'Are you a serviceman?'

'Yes,' I replied, 'I am in the Navy.'

After some time she approached me saying 'You mean the Merchant Navy.' 'No,' I replied, 'the Royal Navy,' producing my leave ticket to prove my identity. She then served me but I could see her talking to another and glancing my way, so I thought it better to eat up and go before the suspicions my appearance presented got reported to the military redcaps, who might not be satisfied by my pass and keep me until they checked my ticket and story, so I took my place, waiting in the queue for the train, arriving at Cambridge late in the evening after all the buses had gone and with no money to afford a taxi, the porter advising me to go to the R.T.O.'s Office. The uniformed man in charge barked 'What do you want?', eyeing me in a doubtful manner. I asked for transport to my village some eight miles out. His curt reply was that he couldn't help. What transport he had was out. So

I said I would go to the police station. 'That would be the best place for you,' he replied, with a touch of sarcasm in his tone. Some fifteen minutes walk to St Andrew's Street Police Station, where I told my distress to a burly police sergeant who lent a sympathetic ear, and without any questions, immediately ordered out a patrol car and two constables, giving them orders to take me to my home. 'You have a shipwrecked sailor, so look after him,' was the obliging sergeant's remark as he held the door open for me to sit in the back seat and off we swept through Cambridge on to the Newmarket Road. Turning to me the constables wished to hear my story. I told them a little about my ship. Stopping, they asked if I minded them giving a lift to an army officer who was waiting by the roadside. 'Let him get in,' was my reply, the constables saying to him as he opened the door, 'Take a seat. Here you have a companion who has seen action and can tell you something of the war.' 'Thank you,' said the young army captain, taking his seat beside me and turning to address me. 'So you have been in action?' he asked.

'Yes Sir,' I answered.

'I shall have my men over the Channel in three weeks' time. We are anxious to get overseas and get on with it,' he told me in his polished voice.

'And you will soon be back. You won't stand a chance against Jerry,' speaking my mind rather bluntly.

'Oh yes we will,' he contradicted. 'It's just a matter of time.'

'You'll be back quicker than you went. Jerry planes everywhere,' I told him.

'I don't think we will,' he answered.

'Well,' I said, 'we have just brought a thousand troops from Norway who didn't seem to know what they had to do when they landed. Lucky for them they had the sense to return right away, otherwise they would all have been killed or taken prisoner by now.'

'Oh, I am sure we will be all right,' were his final words, as the patrol car pulled up for him to alight.

'You'll see when you get there,' were my parting words. Then turning to the policemen I said 'He doesn't understand. We don't seem to have anything to stand up to the Germans with yet.'

'Well, these soldiers have been hanging around for six months, getting bored waiting to go overseas.'

Then I remained silent, for the patrol car was passing the old village pub. I alighted, thanking them for their help, and the car sped away. I turned to walk through the door where my grey-haired mother only a few months before had waved me goodbye.

I stood in the pew of our village church as the bearers walked slowly down the aisle, carrying the coffin of my dear mother, resting it gently down, the family drawn together in grief, standing with bowed heads in the foremost pews, the vicar announcing 'We will now sing the hymn – "There is a green hill far away" - that was her favourite hymn'. Then we all knelt as the vicar read the prayers, saying a few kind words about my mother in his service. Then the bearers came, lifted the coffin up and we followed behind in twos round the old church. (A few village onlookers were staring in idle curiosity; it was the family's private grief and I resented them looking on.) Stopping by the grave, they lowered my mother to her last rest, the vicar saying 'dust to dust, ashes to ashes', and it was all over. Turning away to walk home, my father having to be supported by his daughters to enter the home where we all knew but never said that never again would family life be the same, for love radiating from our mother, holding the family together, had gone. We would fend for ourselves as separate families in the years ahead.

CHAPTER TWENTY-SEVEN

Next day I caught the train to Gillingham, leaving my young sister to look after my father and keep the home going. I was welcomed by my darling wife; our young son gurgled and laughed, bringing fresh hope in the few days left of my compassionate leave, before returning to my Scottish mansion.

There for the next few weeks I continued my life of a lord and gentleman. A round of golf or a walk up the steep hill to the Loch Tom. Continuous days of sunshine from a clear blue sky. No convalescence could have been more bracing, no weather so perfect. As the month came round, from the wireless news and daily papers, I read of the total withdrawal of troops from Norway, with the minimum of loss under the guns of the Royal Navy.

But now the long-awaited offensive of the German Army began on the 10th of May. We all knew how bogged down both sides had been in World War I, slugging it out in rain and mud. Holland could flood her country, the French had built the Maginot line, an impassable fortified network of underground defences. Our troops had six months preparation to plan and build up their positions. At last a stand against that dominant figurehead of Europe, so long having it all his own way. Now we would see what this little strutting gamecock could do. There would be numerous tanks in this war. In the last they had just made their appearance, but the main uncertainty was in the air. Could these flying machines change the whole pattern of war? For years Britain had sent her armies overseas to fight her battles, built up her navy second to none, keeping the foe from her shore, but a taste of the Zeppelin over London dropping small bombs had left a dreadful fear. For I could remember the country folk in my village stopping to talk after the celebration of armistice, saying 'Ah, this is not the end. They intend to come back. They have found a way over the top of our Army and Navy,' as they recalled the night the Zeppelin passed over, how their windows had to be covered over so as not to show a chink of light. The vigilant village exempt class had formed patrols and my poor old Aunt had been summoned by the district magistrate and fined a pound for showing a light from a tear in the bedroom blackout, when she went up the rickety old stairs to bed carrying her lighted candlestick, even though her brother had died in France defending Britain.

Churchill the warhorse had been hammering at the door warning of danger. At last after many years the door had been opened and he was now First Lord of the Admiralty. A cheer had gone up from the rank and file of the Navy when this had been announced. At last he was in harness, and the people now had to face the enemy, who had armed and trained to cast off the humiliation of having to sign peace terms, surrender their Navy and lose their overseas colonies.

While I enjoyed the peacefulness of the Scottish mansion, the whole Western battle front erupted. The French wondered what had hit them. Belgium and Holland were helpless in dealing with the many paratroops dropped on their airfields and strategic points, giving the Dutch no time to flood their land. All was being swept aside, as the advancing Germans swept onwards, with tanks drawing up at garages to be filled up with petrol.

What ordinary human being could understand what was happening or keep up with the rapid advance? It was bewildering and beyond understanding even to those who were commanding the forces against this blitzing, the alarming news appearing in our papers undermining the confidence we had in Chamberlain's government, with him holding his piece of paper declaring peace in our time.

In a matter of days, Holland and Belgium surrendered. France was rearing backwards and our own British force was being pushed into the sea.

In Parliament Chamberlain resigned and Winston Churchill succeeded him. At last the old warhorse was in full command, bringing a feeling of relief to the British people, who now began to realise they had to dig in, get ready to bear the onslaught that was about to be thrust upon them, for the six-month build-up of equipment for the British force had all been lost, its men having to be rescued by a fleet of naval vessels, assisted by the combined fishermen and yachting enthusiasts. Now it began to dawn on the ordinary folk that we would be all alone defending our island.

It was during this time that I was transferred from my convalescent mansion to the Naval Hospital at Gillingham and there given sick leave, now aware of all the drama that was going on around our coast, for it was to my village with wife and son to spend our leave, knowing it would be back to sea as soon as I was fit, but the fast-moving onslaught of the German advance was so devastating and incomprehensible to us all, especially the old soldiers of the First World War, who had stood for months in mud and water, slogging it out. The French towns and districts they had known so well were now overrun and captured in hours. Now the wireless and papers were warning us of fifth columnists in our midst: posters stared at us, warning of enemy spies who could be listening to our gossip. In our village pub, where people assembled and talked of the disheartening news, setbacks and retreating, the arrows pointing on the map of the daily papers, of the forward advances of the Nazis, the smaller and smaller space of our hemmed-in British force, then the arrival home of one or two of the village soldier lads, to tell of the might and well-trained forces of the Hun carrying everything before them, the sky black with their planes, dropping paratroops, tanks carrying infantry along, the dive bombers peeling off to shatter the strongest nerve with their whining terrorising screams to drop their bombs and then to turn and machine-gun the whole population, for no one was allowed to hold up the troops of the Fatherland. The age of chivalry was over, the front line was the open towns, houses and factories were the target for bombs and shells, not fortified enplacements: they could go round those.

Rapidly the British people seemed to accept or understand this, making them more determined to dig in. Resentment against our leaders and commanders of our forces ran high, but even these were not altogether to blame. In this hour of crisis the whole nation as one knew we had got to stand firm. The man who had warned us for many years, who had cried in the wilderness, was now in charge, leader of the country; in him we had trust and faith, and while the village people watched through the curtained windows at a car stopping too long, keeping an eye on a stranger, ready to report if the fifth columnist operated here, they went about their work aware that every grain of corn would now be needed, time to rebuild our out-of-date weapons, retrain our forces. We didn't even know defeat; we had been driven back into the sea, but the sea was our friend, for centuries it had washed our shores and we had learned to ride its waves, a lifebelt of safety.

So that day when I travelled back by train to my depot and a lady got in the carriage saying France had fallen, no panic appeared on our faces, just a calm calculated assessment of the situation was weighed up in our minds. The strength of the British character made us set about our task of standing firm to hit back when the opportunity presented itself.

'Weren't you recalled?', was the question put to me as I entered the main gate of Chatham barracks, presenting my sick leave ticket.

'No,' was my answer.

I now joined the many thousands of sailors in the large blocks where the training of recruits, the drafting to many ships was going on, occasionally running into an old shipmate, hearing each other's stories of our escapes from gunfire, bombs, mines and the hazards of the sea, for almost all had now taken part in some kind of action. Veterans of this war, which in a few months of waiting and only a few days of land warfare was now on the doorstep of our homeland.

What would happen to me? How long could I remain in barracks, a native of the Medway towns, able to get home three nights out of four? My badge acclaimed me a gunnery rating. We formed up on the parade ground with rifle, belt, gaiters and wearing tin helmets, always at the double as we crossed the boundary line into the precincts of the gunnery school. With fixed bayonets we rushed at stuffed figure-shaped bags, marched through the streets of Gillingham, on to the Darland Banks, a steep incline, to toughen up our slackened muscles and straighten up from our rolling sea gait, as in sections we fought the imaginary German invader. The march back through the town, with the band playing to keep us in step, the fruit growers on the outskirts of the town running eagerly and happily to hand us an apple when marching by, almost cheering us now, when a few months back they shunned the happy-go-lucky sailor, almost locking their daughters up to keep them away from jolly Jack.

I relished my stay in barracks, three nights ashore out of four, being trained as a soldier. Was I fit from my ordeal? I thought so but one day I collapsed, rushing up the Darland Banks and was told to report to sick bay – which I didn't. Then one afternoon when proceeding to the main gate on night leave, the sirens wailed their warning. Not much in that; now getting more regular, but on looking up into the sky, hearing the drone of many planes and from a break in the clouds, a large formation of enemy bombers appeared, roaring their defiance but banking as if they were being turned back. I stood petrified in the few seconds these were visible, than I rushed with others to the nearest air-raid shelter to await the all clear. Not long before we could resume on our way, but wasn't I trembling, realising my wife and son were in the firing line. They must go to my village home and I must pull myself together or I would become a nervous shell-shocked wreck. Arriving home to relate our new danger I watched the others' faces in real alarm. What could we do, for the Germans were going to rain down bombs on our homes? Then in the evening a battery of guns opened fire only a few hundred yards away. Shrapnel rattled on the tiles of the roofs. What could these poor defenceless people do? Just sit and wait to hear the terrifying drone approaching, that sinking feeling in the pit of the stomach, waiting for the thump, thump. I trembled at the thought of their fate and showed more distress than the rest of the family, although I tried to be cool, calm and collected. On shore you had to sit and take it, on a ship you could hit back.

A taste of what was to come, for the Germans were testing our defences, preparing to take this little island, which now the high Nazi war Generals must have considered as just a thorn in their side. We would be annihilated. A few weeks to gather their forces together, then the Luftwaffe would swamp us with their deluge of bombs. It was only a step across to England.

But the British people rallied closer together – fear and dread, yes, but I don't think anybody thought we would be defeated as they set about their daily task, preparing to meet the threat by their fire-watching duties, with saving and storing for the coming years of hardship.

Now my stay in depot wasn't going to be for long. Daily I looked at the notice board in the drill shed, reading the many pages of names which were posted up, drafting men to all the ships that were to be manned, and behold my name and number appeared, and to which ship? None other than my old ship again H.M.S. *Eskimo*, which I learned was being refitted in one of the many shipyards somewhere in England. With this startling information

I broke the news to my wife, saying 'It's back to sea again in my old ship. It's being patched up and there doesn't seem to be anything I can do about it. If there's going to be an invasion I would rather stay here,' talking this over as the siren wailed its warning, which was now becoming a frequent occurrence, 'and you', I told her, 'must go back to my village out of this danger area, for the sake of our little son'.

Next morning, turning over all the events in my mind, of my old shipmates, would it be the same, living again in the very same messdeck where my messmates had been blown to pieces? What must I do but put a request in to be taken off draft. Now the drafting commander must have been a very busy man, but still had time to listen to complaints and compassionate requests that were handed in by many ratings for some specific reason, so I was taken before the commander, ready to say my piece.

'Well, why do you want to be taken off draft?' he asked, as I stood before him at attention.

'Sir, I served on the *Eskimo* when she was in battle and all the for'ard messdeck got badly damaged and nearly all my messmates either killed or wounded, and to live again on the old ship would bring back haunting memories, if I served again in the same ship, living on the same messdeck Sir,' I concluded, having rehearsed it over in my mind, trying to make it sound effective.

'Well now', the commander, speaking slowly, looking me over with a kindly eye. 'This is not so in many cases. I have lots of volunteers requesting to go back to serve in their old ship again,' speaking not in an official tone but in a most friendly voice, asking me a few questions about what took place. 'I feel sure you will settle in and keep up the fine performance of naval traditions which the *Eskimo* has so magnificently carried out.' Hardly giving him time to finish I heard myself saying 'I will go back, Sir', my enthusiasm to face the sea again overcoming my desire to stay in barracks, where I would be close to my wife and child in case of an invasion.

CHAPTER TWENTY-EIGHT

Rejoining the *Eskimo* at Barrow-in-Furness, a new ship's company, and a partly new ship. A splendid job had been carried out in the shipyard and she looked again like one of the fleet's dependable ocean watchdogs, all ready to take up her duties, admired by those who had refitted her, for it was unbelievable that a ship so damaged could be towed across the North Sea. Not until she was in dry dock could some of her damaged compartments be pumped dry and the remaining bodies recovered, and this had deepened the sympathy of the people of Barrow, and they gave the ship's company a most welcome reception, with free access to all places of entertainment; from the dockyard worker it was 'have a drink, Jack', as soon as one entered the dockyard tavern. Those few days of enjoying the hospitality of the people of Barrow-in-Furness, and then out to sea, working up to a high standard of efficiency, and the *Eskimo* was back, taking her rightful place with the Fleet in a first degree of readiness.

While our South-East corner fought its aerial battles over the Channel, we guarded the convoys patrolling the sea, ready to be thrown in to repel the Nazi invasion if it was ever attempted. But owing to our daring fighter pilots of the Royal Air Force, the invasion was not attempted. Churchill thanked the airmen with 'never in the field of human conflict has so much been owed by so many to so few'. It was his speeches in the dark days of our hammering that gave us cheer and kept our spirits up, and he did it not by fierce threats, but simple words with so much feeling that they touched the hearts of all. 'We will defend our island whatever the cost. We will fight on the beaches, we will fight on the landing grounds, in the fields, in the streets and hills. We shall never surrender.' Our Hurricanes and Spitfires took off, sailing in amongst the Nazi bombers, just as Drake had smashed and battered the great Spanish Armada with his small manoeuvrable ships.

Many times did I glue my ear to catch the faint wireless news during those dark days and months as we brought the convoys through, to hear how many German planes had been brought down each day. This cheerful news heartened me as I searched the horizon for any kind of danger. Perhaps the concentration of so much activity on our South-East corner left our sea lanes alone for the time being, for we saw no further action than an occasional dropping of a depth charge by the *Eskimo*, as the new ship's company settled down, keeping watchful in the many sweeps with the fleet over the North Sea and right up into the Arctic.

But now a new wave of young men were coming into the service of our Navy, fresh from schools and universities, men with only weeks of training when we had years of experience. Some of these were young intellectuals, who did their six months sea time before passing out as sub-lieutenants. These young men were fair game for some of the older sailors, who now saw the opportunity of getting their own back on the future wardroom members. But these young men soon overcame this crude lower deck taunting, to become worthy sailors in theory, if not in practical experience.

I enjoyed their company and many a friendship developed as we stood on watch. I voiced my opinion on religion, politics, poverty, wealth and sundry other topics, often wondering what they would think, when as officers they would look back on their first twelve months of naval life, recalling cooking, scrubbing, cleaning, or scraping the paintwork, being kicked around by all until they gained their sea legs. Relieving one of these ratings from his hourly turn as lookout, he related to me that the First Lieutenant had reprimanded him for his blindness in not reporting, saying 'Lookout, why didn't you report that ship ahead?' He had answered back in his cultured voice 'It is so obvious, Sir, that I didn't think it needed reporting.' It was our turn to worry if our homes were hit by bombs, our families killed or maimed, but the British people had earned their respite. Time somehow would be on our side, although when one looked at maps to see almost the whole of Europe occupied by the Germans, from Norway to the Bay of Biscay, one wondered how we were going to survive.

Once again I left my ship to travel down from Scotland to the hub of activity in the barracks. Why I didn't know or couldn't guess, but it meant heaven to me to be back home enjoying a few days leave with all the heaviest of raids over, with not much fear of the Germans risking a daylight raid.

The Medway people seemed to be breathing more freely again. During those critical months, my wife had taken our little son to my village, but was now back, and brimming over with delight told me all about the job she and her sister had taken, helping in a large nursery and market garden close by. Long lines of elderly people and young children queued up on Saturday mornings for vegetables, she told me, enabling young mothers to keep at work. She also reaped the benefit of being allowed to buy vegetables at a low price, and every Saturday morning filled a large basket.

A few days leave, a couple of weeks in barracks to enjoy and then my name and number appeared on that same old notice board in the drill shed, but this time it was under a number instead of a ship's name. This led to many speculations, but however hard I tried, I couldn't find anything out about this numbered draft, only that a large number of names were posted up.

The following morning I stood on the platform of Chatham Station with the remainder of the draft, taking a train to Liverpool. But it wasn't a ship in the Liverpool docks we were marched to but on to the ferry and over to Ireland. What was in store for us, we wondered, as we loaded our bags and hammocks on to the waiting lorry, looking untidy after enduring the uncomfortable crossing. We formed up in some kind of order to half walk and half march, as the rope ends and wire hawsers lying loose on the dockside had to be avoided. This led us to carry on talking in a spirited way, wondering where our journey was taking us and what would be our destination. We were staring curiously at the squat shapes and sizes of the many merchant ships we passed, but on rounding a bend, a short distance ahead, there stood the lorry with our bags and hammocks and one or two figures in naval uniform waiting. 'This is it,' we declared, and were called to a halt opposite a fairly large merchantman. A Petty Officer came over to tell us this was our ship and to carry aboard our gear.

I looked up with delight, feeling a sudden surge of pleasure. At last I would know the feeling of sailing along at a steady, unhurried speed, having time to stand relaxed by the deck rail, watching the sea roll by. How often I had observed the merchant sailor, watching us in his leisured time. Now I would be doing the same, seeing the slender destroyers dipping in the swell, gathering speed to round up the straggling convoy, for these squat, broad-beamed ships steamed so leisurely along, with only a steady roll once in a while. No feeling sick on these sort I was thinking, stepping aboard its deck. Quickly we stowed our gear, sorted out our messes and walked around to become accustomed to our new ship.

This numbered draft was the merchant ship *Springbank* of about 11,000 tons, its top speed being eleven knots. It was equipped with a battery of four twin-barrel 4″ guns for repelling aircraft as its main armament, and also had modern, smaller close-range weapons. It also had a catapult. No doubt we would soon find out more, but the loudspeaker was calling the ship's company to muster where we would be told to which part of ship we belonged, assigned watches and so on. For I was now a Leading Hand, wearing an anchor on my sleeve, having been promoted to a Leading Seaman before leaving barracks, my promotion being entirely due to the rapid expansion of the war. In peacetime I would never have attained a higher rating in my short service without signing on for a longer period, but now I stood facing the ranks of these young men, nearly all now joining since the war started, and I must say I felt rather elated at the feeling my position gave me. I would now have the responsibility of the discipline of each member of my mess, seeing that they understood and carried out their duties, some of which must have been very irksome after civilian life.

It was now eight years since I had left my country life to wear bell bottoms and a blue collar and sail the sea. Now here before me sat other young recruits who only a few weeks ago had left their homes and families to be flung into war, looking up to me for guidance and advice. What had I got to tell them or show them? for wasn't this some kind of Q ship, to be disguised for some special reason? I had heard that U-boats were hunting far out in the Atlantic owing to their increasing range and taking heavy toll of our convoys. Even in this war I was taking part in, I understood very little of the strategic pattern, which covered the whole of the Atlantic, when new devices were being put into operation and kept secret as long as possible, before the other side could counteract them. We had kept the U-boat at bay by our ASDIC detecting gear, the acoustic and magnetic mine had been countered while radar was still secret, but we as ordinary sailors understood it had given valuable assistance to our air force in winning the air battles, and now was assisting us at sea so that we could pick up objects at night or in fog. It was like seeing over the horizon and having those few minutes to get prepared. What had the Germans brought out to cause our Admiralty to take a valuable cargo vessel, and give it modern equipment for repelling aircraft, ASDIC and radar sets aboard with that catapult as yet empty? Would we remain based in Belfast where numerous convoys converged to feed our large Western ports?

Belfast absorbed our ship's company with night leave for the week or two it took for the dockyard to put our ship in order with us becoming settled in, working up into a fighting unit. But what pleased me most, and gave me enormous pleasure, was my special job as Coxswain of the *Springbank*'s motor boat, sharing this special duty with another Leading Seaman. The old ship only carried one power boat and it was hoisted out and in by steam power, a single wooden beam going fore and aft of the boat, shackled on to the bow and stern, a wire sling attached to the derrick wire, which took the weight round a steam-powered pulley, easily hoisting us in or out as required. This meant that whenever we dropped anchor and the motor boat was called away, my crew and I would be ready in the rig of the day, at the first pipe of 'Away motor boat', when it would be hoisted out, unshackled and away – taking the Captain ashore, picking up the mail, taking off and picking up libertymen, my crew consisting of a bow and stern man, with a stoker as engine driver, with myself taking the wheel, giving the orders to the stoker, who sat at the engine just for'ard of my feet, and to the bow and sternman as I manoeuvred to come alongside.

When the old tub rolled out of the shipyard, she actually got away on her own engines, with Lofty Day saying he didn't know whether she was moving or drifting, but as there was smoke coming from the funnel and a slight movement of her deck, a quiet cheer went up from the seamen working on the upper deck. To think we were back at sea, almost ready for combat operations again, although we wondered what our duty would be at the slow movement of this lovable old tub. We would anchor in the bay of Bangor, some distance out from the small pier of that little holiday resort after steaming around, with the engineers putting right the defects and the officers getting the feel of the ship. We anchored in the lovely bay, and soon I was at the wheel of the motor boat, ferrying a boatload of some forty or fifty libertymen, turning to ride in on the swell, feeling as proud as any large liner's skipper, standing erect at the wheel of my small craft, my blue collar flapping. It was the one time in my seafaring life that I felt the pull of the sea, its true call, like those old salts whose love of the sea is written in their weatherbeaten faces and shows in their eyes. For just as I had looked to the senior hands when I was a young ordinary seaman, so did these young recruits look up to me.

'Now you youngsters are given a sailor's uniform, just sent aboard. It's a good job old Hitler doesn't know the true state of the Royal Navy,' I would tell them with a twinkle in my eye, shaking my head with doubtful concern at the pitiful expression on their guilty faces, after coming off night leave and hungrily sitting down at the breakfast table. 'I hope you have washed those dirty hands before touching that bread,' I would say with a withering look, adding 'how does one know what your hands have messed around with after being ashore all night. Cleanliness is next to godliness, and I'm going to keep it that way.'

But much too soon for us our old tub became more lifelife and her steady movement a little more restive and we became aware our stay around Belfast and Bangor was about to end, for on board to join our ship's company came members of our youngest service, a sergeant pilot and one or two other members of the R.A.F., and above our catapult a large crane was lowering a fighter plane, with our Commander strutting around, an air of knowing a secret stamped on his face, looking up with his nautical eye, asking the R.A.F personnel questions, as if in doubt at their ability to ship a plane aboard and secure it on to the catapult, but after some little time there it was. We all watched the R.A.F. crew adjusting the wire stays with the sergeant pilot sitting in the cockpit, getting the feel of his controls.

That dinner time we looked across at the mess set aside on our spacious ship for the R.A.F. personnel, talking over between us the new addition to our ship. 'It will make us safe from air attack,' was the main opinion. 'Give Jerry a fright to meet up with a fighter in the middle of the Atlantic.' But we were soon to have our minds put at rest and the role we were to play revealed, for as soon as our fighter plane was aboard, our dinner over, the old *Springbank* got under way to anchor in the bay of Bangor.

Already some were bemoaning the fact that they were always at sea, never getting any shore leave; while others in no. 7 mess were scribbling away, getting off their last letters to wives and sweethearts. Myself, sitting waiting, wearing a sou'wester, with the remainder of my oilskins ready to slip on, for it was wet and murky out in the bay. Soon the call 'Away motor boat', making our last contact ashore before the old *Springbank* set sail. I quickly donned my oilskins and soon we were speeding away through the lashing rain with the last censored mail and orders to await the postman and an officer. Soon we were racing back, my eyes searching through the gloom, keeping a straight course to the anchored ship, just discernible through the gathering gloomy mist, hoping the rain would hide our departure from the watching eyes of the spies ashore.

Down to the mess, to be informed the Commander was going to address the ship's company shortly and we were all to muster for'ard. Long before the actual pipe to muster, nearly all were assembled, curious and anxious to know what he was going to say.

'Now,' commenced the Commander, 'you have all seen the plane we have aboard and probably wonder what role we are going to play. The war at sea is vital to our country's existence and all could be lost if the food ships don't get through. So far we have held our own against all the Germans have launched at us but it's a tough battle against the U-boat. The slow-moving food convoys are so vulnerable to the attacks of U-boats, which are getting more daring and hunting in packs, while their long-range Focke-Wulf planes are searching the ocean, finding our convoys, plotting their direction and speed and calling up all the U-boats in the surrounding area. The *Springbank* is to position itself in the centre of the convoy, giving air protection. If we can bring one of their planes down by gunfire so much the better, but as a last resort our fighter plane will be catapulted off to attack and attempt to shoot it down. Now this is almost a suicidal attempt for our pilot, for if there is no friendly land near for the pilot to land, he will have to return – that is if he can find the convoy again – after his chase and come down by parachute in the sea, with the chance of being picked up by one of the escorts.

'So much depends on the efforts of every one of you to make this a success. The Admiralty have given much thought and valuable time to convert this ship to combat the new danger facing our convoys, and I must convey to you the need for the utmost caution and silence when ashore. Do not say a word or drop a hint of what you are doing or the type of ship you are on, for an unguarded word may lead to the loss of ships and lives. Your mothers, wives, and sweethearts depend on us. Don't let them down.'

With the last words ringing in our ears, the Commander stepped down and vanished through the watertight door of the ship's after-compartment, leaving us all rather silent and reluctant to move.

I went to check my motor boat, to make sure it was safe and secured, lingering a few minutes to look around. Darkness was gathering, only a few navigation lights flickering through the night. The quietness was broken by the ship's broadcasting for special sea dutymen to close up, the hiss of steam as they put on the capstan, the clanging noise of the cable drawn over the deck and down into the hold, for the old tub was slowly moving, stealing out into the night.

Next morning I looked through the binoculars of my training sight to see a ship, then two and three, with a larger number becoming visible, being shepherded by the small corvettes. No doubt we wee a welcome comfort to these merchant ships. Many of the crews stared hard, with eyes riveted on our plane as the *Springbank* took up its position in the centre of the numerous merchantmen on a clear, bright morning when I could clearly see the little corvettes far out, ringed round the convoy with here and there a destroyer, its long slender body almost dipping under the water and only recognisable by the white wake it was churning up as it outraced the slower flat-bottomed corvettes. The convoy zigzagged into the Atlantic at a very slow speed, turning together and keeping station. Although the zigzag was designed to offer a difficult target to a U-boat firing a torpedo, it also became a hazardous manoeuvre if not carried out by all in the correct time, so a constant lookout by the navigating officer in keeping proper station was essential.

I was enjoying this cruise, walking up and down the deck of this spacious vessel as it proceeded at a steady ten knots; even in a half gale the old tub just about rocked, and even when it was rough you could remain standing on the upper deck, letting the salt spray rain against your oilskin and watching the snarling sea dashing and breaking on the old *Springbank*'s steel sides. She only bucked occasionally or heeled over slightly when a large wave caught her broadside on. As the days passed by the crew of naval matelots became fond of the old tub, especially after seeing our first convoy out and bringing another safely in through the dangerous U-boat-infested lanes. Nothing troubled us. Occasionally we heard the rumble of an exploding depth charge dropped by a destroyer on the horizon, and the only long-range plane was the R.A.F. Sunderland, a four-engined seaplane whose welcome appearance some thousand miles out in the Atlantic gave us the feeling we were in close contact with those at home.

Then one morning we left our straggling convoy to drop anchor in Bangor. In the rig of the day, I mustered my crew standing ready in my motor boat to be hoisted out. Then we were away, plying to and fro; then later on we went alongside in Belfast, where we could resume old acquaintances. In the daily papers we read of our setbacks in the Middle East, and of how the Nazi storm troopers had invaded Yugoslavia and Greece, sweeping all before them. Night bombers were still giving our towns a bashing and Hitler had boasted that in the Battle of the Atlantic U-boat and aerial activity would be intensified. The all-round pressure by the Germans against the British defensive positions was crippling. Churchill was having to explain in Parliament our withdrawal in Greece, our towns were being burned and our scanty food rations were being cut. It was as if a black cloud overhung our war effort.

We had seen nothing of our enemy during our first trip. The young seamen said all they had seen were numerous ships slowly proceeding under escort, while they lazed around a gun, when at home they had worked in a factory, turning out aircraft or doing some kind of work to help the country's war effort. Now, they told me, they felt their life was idly wasted, being led to believe their sailor and soldier heroes lived in hourly danger of being blown to bits. They said that their old Dad's fire-watching duties were more dangerous than sailing the sea.

'Don't get too bloodthirsty,' I said, 'longing to send German sailors to a watery grave. Have patience. I only hope that nothing does happen, but if it does come it will be swift and sudden.' But this didn't help when our second voyage was as free from action, as if we were sailing on some inland lake. The Sunderland seaplane was joined by a Catalina monoplane which flew above us during the daylight hours, while we strolled up and down the deck of the old *Springbank*, having an occasional practice run during the long days and nights, feeling secure in the middle of the company of so many ships. The summer months passed by, with the *Springbank* wearing her engines out in the many sea miles of sailing, which landed us in one of the Belfast dry docks, with ten days' leave being given.

What a joyous messdeck, the busy preparations with witty remarks and funny jokes, snapping closed one small suitcase, collecting our leave and railway vouchers, then away, pouring off the ship, and hurrying to the ferry to Liverpool Station, the excitement to feel the rattling jolt of the railway carriages. No word of warning can be given to your wife or others to know you are on your way. You just arrive out of the blue, dirty and dishevelled from the long train journey. At last I opened the garden gate, stepping light-heartedly down the back path, to open the door to the startled exclamation of a joyous reunion.

'It's you who have been in the front line. I have been cruising on the sea in a spacious liner, sailing along so leisurely because it is so big that the largest wave only gently shakes our faithful old tub. I've forgotten what it is like to feel seasick.' I assured my wife that there was no need to worry, for the roughest trip was the ferry over from Belfast to Liverpool and the only danger I faced was of getting trampled underfoot in a rush to get a seat on the train to bring me home.

'And you must forget all about those cabbages, lettuce and tomatoes which you attend and tell your boss your sailor husband is home and demands your full attention for ten whole days.'

'Yes, my Admiral,' she murmured with that coy look in her eyes, 'your command will be carried out without delay,' stepping from my side where she had led me to see our growing son, sleeping soundly. In a second she had her coat on, and grabbing my hand rushed me off to the place of her employment to tell her boss she wouldn't be in to work for a fortnight. For I had to learn to be a father to my young son, now over a year old, walking and talking.

The dread feeling of my leave drawing to an end marred our last few days, however much we tried to keep the conversation happy and away from war and my fast-ending leave. Longer minutes we lay in each other's arms, saying hardly a word as the fatal day drew close. Then the last family outing together down by the river, and that evening I took my little son up to his bed, tucking him in for the last time.

I walked aboard the *Springbank* to be told it would be another fortnight before we were ready for sea. All remaining hands had been sent on second leave, the ship feeling almost deserted during the day. I busied myself cleaning and painting the motor boat, but during the evening after some had gone on night leave it only left a bare number, with heaps of time to sit down and talk of our leave and the events of the war. There was Ordinary Seaman James Peck, who came from the Midlands, and felt proud of himself in his sailor uniform, who related to his shipmates that he had told them all at home every time we had gone out we had sunk a U-boat.

'Why tell them that?', said Jack, 'you know we haven't seen a thing.'

'I had to,' replied young Peck. 'They kept asking – and anyway, what was that destroyer doing dropping depth charges on the outside ring of corvettes? Often during the night there was a rattle and thump as if something had been sent to a watery grave, so I don't think I was telling a lie. Anyway, it's better for their morale if you tell them something. My Dad was real proud of me, taking me along to his club. We all sat round the table, him and his chums, listening to me telling them I was on a mystery ship. I couldn't reveal anything of what we did but that we went out with convoys and hadn't lost a single ship.'

'Might as well have a good time and spend out, for the war looks as though it's going on for years now that the Germans are advancing into Russia,' said Jock later on. 'What do you think, Hookey? Will the Germans beat the Russians?' They referred to me as Hookey, on account of the anchor on my Leading Seaman badge.

'Hitler has certainly taken on something attacking Russia. 'It's giving us a break, seeing the Germans use their strength against 150,000,000 Russians. Thank goodness Hitler has thrown all his might at them, it will give us time to concentrate on winning the war at sea, keeping the Middle East army supplied and, of course, getting our food ships through, which is giving the admirals a lot of thinking,' shaking my head as if in doubt at their ability. 'It's getting mighty serious at home, rations becoming less with hardly anything else to be had in the shops. It's tough for our women hunting in the shops, finding something to last the week out, trying their best to hide it from us.

'We have got to keep this old tub going somehow because I've a feeling that this war at sea isn't altogether going our way. It might get worse, and our next voyage on this old ship may be quite different, with young Peck here having a real tale to spin round the table of his father's club. For all I know they may be fitting us out to go further out into the Atlantic or could be the Mediterranean.'

'Not fast enough for the Med.,' declared Jock. 'Need fast ships to get through to Malta.'

'I don't know; if we haven't the speed we have got our guns,' I argued, 'but don't you think we had better turn in and get some sleep?'

'It's a bit late for old Jerry to bomb Belfast tonight,' remarked Bill Murray the young A.B., getting up to sling his hammock. 'But we might as well get a night's sleep, won't be long now before we are cooped up all night.' For often we sat up talking well into the night when restrictions were relaxed with only a few men on aboard, but these late-night discussions soon ceased when the full ship's company returned.

Once again we heard the message: 'The Captain will address the ship's company. Clear the lower deck, all hands muster for'ard.'

'I hope you all had a nice leave and found your people doing their best under the trying wartime conditions. While we have been on leave, the dockyard workmen have been very busy making us seaworthy and overhauling the engines. So far we have been very lucky in not encountering any attacks or losing any of our convoyed ships. Perhaps some of you who have recently joined may wonder if all you have heard and read about ships being sunk is just an old sailor's yarn. I can assure you *it is not* and the war at sea is at its highest peril at this very moment. There are U-boats lurking out there in larger numbers than ever before, and at this critical stage we have got to do something about it. In the last few months you have all endured wartime conditions at sea and are thoroughly trained to meet any attack our enemy may make against us. I know you are all eager to strike a blow in retaliation for all the atrocities our ships have had to face up to. You are now being called upon to carry out a most dangerous and hazardous voyage, where our enemy has been 100% successful. In the last two convoys not one merchant ship has escaped the attack of the U-boats.

'The last two convoys to leave Gibraltar have been sunk. We have been given the task of bringing through the third. With our 4″ guns we hope to spring a surprise and shoot the enemy's long-range plane down before it can call up the U-boats; as a last resort our pilot here will be catapulted off in an attempt to bring it down. By your devotion to duty and everyone's utmost effort, don't let us fail, for there is a fortnight's leave awaiting you on our successful arrival back.'

With this last announcement the Captain stepped down, leaving the Commander to say a few words in thanking us for our hard work in getting the ship looking in good shape after its refit. We resumed our discussion, commenting on this new information, but before we had time to settle it was 'up hook and away'.

We joined up with the convoy, slowly proceeding on our way to Gibraltar, the weather gradually becoming warmer, with no alarm to send us flying into action – not even a false one. Was this lulling us into a sense of security, peacefully steaming along with the dolphins?

The Rock, a sentinel of security as it appeared to me, came at last into view, splashed in sunshine. No wonder us old hands never tired of seeing it, rounding the entrance, tying up alongside the jetty. The convoy safely delivered, we had a few days' rest, with a run ashore to collect all the wonderful unrationed goods that were for sale in the shop windows. I got a big wicker skip and filled it with bananas, oranges, lemons, grapes; all the lovely fruit which was unobtainable in Britain, carefully stowing it away in my motor boat. An exclamation of delight, the faces of all the family would brighten up when they beheld that skipful of fruit, the rationed goods of sugar, raisins and currants. What a glorious welcome with so many scarce and rich things to place on the white tablecloth at teatime! I would be my mother-in-law's wonderful son-in-law. I hadn't forgotten those packets of tea. What a pleasure it was for me to spend my money, carefully stowing all the goods away, covering them up and tying them down, so that no heavy sea would unship them. Besides, the motor boat itself was high up on the upper deck covered and lashed down. Now for a convoy home, which wasn't long in coming. A few days and the *Springbank* pulled out from the backwater into the Straits of Gibraltar, heading for the wide open Atlantic, taking up her position in the centre of the convoyed ships. It was goodbye to our splash of sunshine and now what? It had been such a pleasant voyage to Gib. Surely nothing was going to upset us on our return trip? We had all heard that the German admiralty knew every movement, time of entry and leaving of each ship that arrived and left Gibraltar, having their own team of observers on the Spanish mainland and being well informed by the Spaniards themselves, who now saw the opportunity of getting their Rock back from the British. But all went well, the lads talking of each other's special gifts they had for those at home. We practised casualties so keenly that I actually became one, getting my thumb jammed in the training handle, resulting in my rushing down to the sick bay, my thumbnail hanging partly off, the doc. saying 'Leave it like that. In time another will grow and push it off.' Two days catching the last of the warm weather, beginning to feel secure, when we heard the rapid ring of the alarm. In a second men sprang into readiness; the pointers on the training dial told me this was no false alarm. Those in the director had something in their sights which they were following. Then came the order 'All guns stand by to open fire. Enemy aircraft in sight.' This was it! flashed through my mind. A minute or two of training from stop to stop. I took a look over the top of my sights and there circling some distance away was a large plane which I took to be a Focke-Wulf, its long body gleaming against the blue sky; a little closer and higher, then we could open fire, but one sensed he was being cautious, weighing up the situation and flying round at a safe distance. Was he studying each ship to see what danger there was and would he come closer when he had completed his observations, seeing there was no larger warship than a corvette, with its 3-pounder guns? Patiently we waited, following round with our guns the encircling enemy plane. Had the German crew already calculated our speed and course, transmitting it to the waiting U-boats over the horizon? Had they been informed of our existence, that one of these merchant ships was fitted out with a battery of 4" anti-aircraft guns and a fighter to catapult off? For this was a cat-and-mouse affair. Waiting for the 'mouse' to come closer to get a better look before we dare pounce. For us to open fire before he was in range was giving ourselves away, that could result in sinking the convoy at night. Jerry circulating, keeping out of danger, not risking their planes and lives by being reckless, and so the cat-and-mouse game went on for some time. Perhaps Jerry thought he was safe or the gunnery officer thought he was within range, for time was on Jerry's side.

We opened fire, our shell bursts not appearing too bad for line and distance but not troubling old Jerry too much. He just slightly glided his plane away and lower, continued circling the convoy just out of distance and just over the tops of the waves. We altered our position to have a more open and clear view, opening fire again, without any success, or likely to have. It now became clear we had to play our trump card. The sergeant pilot in his cockpit was revving up, the R.A.F. crew standing by. It wasn't altogether that easy to fly off. The ship had to be in position so as to be clear of others and the German plane required to be where it was hoped it wouldn't observe our fighter being catapulted off. At last the best possible moment, the signal from the bridge, the catapult's charge fired, hurtling our fighter through the air to the cheers of us all. We watched as the pilot straightened out, rising in a curving spiral, the German plane alert, for it veered but then both disappeared. We had the satisfaction that the German plane was being chased away or frightened off, sincerely hoping our R.A.F. pilot had caught up and shot it down, but we never saw our fighter any more or heard the result, and thought maybe he was in range of Gibraltar and would land there.

All that day we took extra care in watching the sea and sky, and sailed on, so far safe and well. The third morning our hopeful doubts were shattered, our fears proving correct, as the alarm bells rang out with the report 'enemy plane in sight', the old Focke-Wulf making its appearance. Vainly we tried to shoot it down, but the wily fox knew just how close to come, keeping just above the sea, weaving happily while we could only watch and wait, seeing our enemy so near, knowing what he was up to yet helpless to do anything about it, 'keeping this up most of the day then leaving to reach home for tea' was the comment passed, with the evening darkness coming much too quickly. I stood watching and fearing the closing night until the convoy ships were just dark shapes, shadowy silhouettes against the bright glow left by the sinking sun, with that too rapidly disappearing, the waves turning black, the cold night surrounding us, the dark shapes of our nearest ships barely discernible as the night grew black and chilly. No longer could I see any distance and only looking straight down from the guard rail could I make out the sea and hear the splash of waves. I wonder what lay out there. Were the U-boats getting ready to launch themselves against us, striking in the pitch blackness and sending these helpless ships to the bottom? I shuddered at the thought, making sure my life jacket was tied securely and puffed up the rubber tyre one, buttoning up my oilskin, then sat myself on the trainer's seat of the gun, pulling my duffle coat round my leg. I would be instantly ready to act, to train my gun if any action started, but thinking it would be more dangerous to our own ships by firing. We must put our trust in the little corvettes in detecting the U-boats and not letting them slip through the protecting ring, as I didn't relish the idea of jumping into the sea. Ten o'clock – eleven o'clock – midnight – into the morning hours. My keyed-up tension was like waiting for the executioner to come along, dreading the moment of his knock. '0100,' our communicating number reported. Some stirred, trying to ease their aching limbs, a slight chance we might get through the night, I thought, resting my head on my training sights in a vain attempt to doze off. Then I heard the buzzing of my training pointers, being the director training round. 'Follow director,' was the quiet, direct order. Something was going on, for suddenly a cry of 'Look.' In the distance the darkness was lit up by star shells or flares fired by the corvette, descending slowly, their spreading arc lighting up a vast area of the sea. Instantly everyone was on their feet, standing, seeing the flares falling, with eyes searching over the lighted surface of the sea until the last touched the water and went out, leaving the night blacker than it was before.

'What's going on out there?' we asked each other, 'and why shoot flares to light up the convoy?'

'It's the only way they have of keeping the U-boat below the surface,' was suggested. The interruption died down but we all remained on our feet, feeling uneasy about the situation which we didn't seem to be able to do anything about. Soon more flares dropped down out of the sky and the distant rumble of depth charges was heard, giving us more cause for dread and fear, for it seemed to me that the wolf pack was on the prowl, snarling all around, and we on the *Springbank* could only watch and wait. Then suddenly on our starboard quarter a red flash lit the darkness. Simultaneously a loud explosion rent the night air and in a few seconds bursting crimson flames leaped into the night sky and in the orange glow, steam and black smoke belched forth and as we stared into the appalling smoke and flames the bows and stern of one of our convoy ships rose, tipped inwardly until perpendicular, the two parts majestically momentarily suspended in the air before plunging to disappear beneath the sea, leaving the waves to roll over where a few minutes before proudly sailed a merchant ship. Unbelievable – a ship so large could be sunk to disappear before one's eyes in a matter of a few minutes. 'They got one,' someone uttered, breaking the silence and bringing us out of the shock of witnessing this horrible disaster, for the sudden explosion and rapid sinking couldn't have given the crew any time to launch a boat or even jump into the sea. Those not blown to pieces must have perished with the ship. We had witnessed the death of a ship, knowing that we too faced that peril at any minute. This was what it was like to be attacked by a U-boat pack. Uneasily we stamped around, looking out into the black sea, our eyes searching over the waves as the corvettes quite frequently now fired their flares to drop all around in a vain attempt to keep the U-boat at bay. But that night five ships were sunk, the waves just washing over a few minutes later the very spot where they had met disaster, leaving just emptiness. At last the welcoming daylight, for there had been no action for some time, the U-boats no doubt withdrawing, getting clear before daylight, their joyous crew full of arrogant confidence now they had been so successful in sending so much tonnage of British shipping to the bottom. There would be iron crosses and promotion awaiting them on their return to the Fatherland. The English couldn't last long at this rate of sinking.

On the *Springbank*, counting the vessels, there were only twenty ships left. We turned to each other with a look that read – would tonight be our night? We had watched these ships disappear under the waves in as many minutes as it had taken years to build. Many of their crew drowned, men like us who had volunteered to serve on the sea. In some home hung a picture, a photo of a smiling man in uniform he proudly wore. After tonight there would be sorrow and sadness to follow the loss of a husband, a son, children left fatherless. Why must it all happen? It's all right reading about it in the papers, hearing it on the news sitting in your home miles away, but when you're right in the centre of operations, fearing for your own life, it is then you realise the horror of it all.

The old Focke-Wulf turned up punctually in the morning, coming over a bit close, no doubt to count the ships and let his compatriots know the results of their combined mission. Right away we opened up, letting him know that we were still there and very much alive, but the plane just turned away to continue his wave hopping and circling the convoy, knowing he was safely out of distance and there was nothing we could do, only pray that something would go wrong, that he would dive into the sea, one of our aircraft carriers would appear, but hour after hour he continued, then left, so as not to be late for tea. What could we do but stamp around waiting with fearful dread the coming hours of darkness? And how quickly this descended. I took my seat at my gun, switching on the lights on my training dial, pulling close my coat to keep out the chilling night air which added to the dismal effect of sitting waiting, wondering and hoping nothing was going to happen. The dread of the long night, when minutes seemed longer than hours. Would the time never pass? At last eleven o'clock. I watched the pointers of the director in my training dial and knew they were sweeping round, looking through their powerful binoculars. Our guns were loaded; in a split second we could be firing, if only we had a target to fire at. Waiting in the darkness. Then the sky lit up, flares are dropping, descending slowly, showing up the waves. It's a bit rougher tonight, which makes it look lonely and sinister out in the black space. The flares die out. I think the wolf pack are howling around earlier tonight, more hungry and daring after their success of the night before. The flash of a hit by a torpedo. The loud report and in the next few minutes I watch the sad end of a sinking ship. The first of the night's attack. The terrifying ordeal that's going on, burning, drowning and sinking; watching one after the other. At long last the dreadful night is over. The U-boats have stopped their attack and all is quiet, the dark outline of the remaining ships becomes visible. We count the number. Seven are missing. A most successful night for the wolf pack, slinking away to lick their lips in a sly, crafty, delighted, satisfied grin, as we on the *Springbank* walk over to the guard rail, looking at the remaining vessels, fearing that it will be their fate tonight, or maybe ours. The morning hours, and we watch the sky above. Without fail Jerry's plane arrives on time, as we jump into action, guns elevating and training as we follow the director. A burst of fire. Old Jerry knows we are still there. Perhaps that is what he came in close for as he quickly glides away back to his wave-hopping and giving his information to the U-boats, congratulating them on topping the number of the night before, settling down to keep up his continuous round of flying. After about an hour or more of watching this plane of our enemy keeping up its round-and-round flight, the Captain could stand it no longer. Drastic action must be taken against the German crew who knew they must be tantalising us hoping we would take some kind of reckless action, for we felt the old *Springbank* altering course, steaming out of its station in an attempt to get a clearer view and perhaps shorten the range, but when you can only do eleven knots, very little manoeuvring can be done at that speed and to get left behind would be fatal, for Jerry always had a U-boat following up astern for stragglers. But we steamed to the starboard side of those left and hoped that a shell burst might be lucky, but it was hopeless to drop a shell at extreme range and it burst just above the waves. The German crew must have laughed at our million-to-one chance, smiling to themselves as they probably said in their signals to the U-boats 'to get the one with the guns on as it's getting quite daring'. After our vain effort we steamed back again into the centre of the convoy, which I thought made it look as though the merchant ships were to be sacrificed to save us. At the regular time in the afternoon, off went Jerry so as not be late for tea. And now we were left alone before the one-sided battle started. Twelve of us gone, thirteen left. Perhaps the corvettes could make a closer circle now there were not so many of us, for as the night closed in, so did the other merchant ships seem to hug closer to us, as if seeking shelter and comfort under our protection. Tonight the sea was getting that much rougher, the wind a little stronger, with a few showery clouds sweeping over. Would it make it better for the U-boat and worse for us or vice versa? I discarded my duffle coat, keeping my oilskin on, thinking if I had to jump into the sea, the oilskin wouldn't hold the water, and I must remember not to struggle, keep calm and let my life jacket and lifebelt – for I wore both – keep me afloat. This would be the third night. Would they attack? Our guns loaded; a last look at the inky sea. The waves were larger, more ruffled. How uninviting the black choppy waves looked. I took my place in my trainer's seat and in my mind prayed for nothing to happen and sat there waiting. So eerie, so quiet. A black cloud hid the stars. Then a spatter of rain. The wind got up, thumped against the gunshield, then went tearing off in a blustering scream as it found its way round, under and over. Even the old *Springbank* rolled slightly as the wind caught it broadside. As the storm passed over the wind quietened down. The sea was less rough and the ship became steady. I wished it had kept up for it might have blotted us out. It was uneasy sitting waiting, wondering, anxious. There was a hope they had run out of torpedoes and nothing would happen.

    Suddenly a depth charge rumbled in the distance. The dread and fear of knowing there was another night of terror to face. The flares lit up the convoy. The battle was on. Would they keep the U-boats out? It must be half an hour gone and nothing had happened. Then a more powerful explosion; a ship was going down. We could all see it clearly in the light of the flames. Then the stern slipped under the sea. Gone for ever. A few minutes after it had disappeared, another was struck and went down. A third followed. They meant business tonight. I thought of the watery grave of the victims as I watched the closing waves rolling over the empty place. Some must have left widows, a mother has lost her son. I made out the outline of the remaining vessels, hugging closer together. Would this be the only attack? That dreadful waiting, fear inside me. Midnight gone, then one o'clock. A sudden order shouted by the communicating number 'alarm port', bursting into activity as the 4" swung round, the twin barrels depressed. We were about to open fire. There was a terrific shattering – and it wasn't the gun firing. The *Springbank* was heaved out of the water. Another shook our faithful old tub. A second or two afterwards, further for'ard. Many feet scampered around for she was listing over. All was confusion in the blackness of the night. I jumped from my tilting gun, found my way along to my 'abandon ship' station, which was number five Carley float. Quickly, with others, we unshipped it from its position, tipped it over the side into the sea, making it fast to the guard rail by its towline, for we had been struck by two torpedoes on the port side and were tilting over to port, thus making the distance not too great to jump in. As yet there was no order to abandon ship and I turned my attention to assisting others. An attempt was being made by an officer and a few ratings to get a small boat over the starboard side, this being at an acute angle, and heaving and struggling we managed to get it over and lowered away. It was then I noticed the tremendous height from the upper deck to the sea on the starboard side and also a big sea was running, this being the weather side and the boat was swinging against the ship's side, so the officer called out for two volunteers in the boat. I looked down, shaking my head, but two volunteers had come forward and were swarming down the side. 'Keep it off the side,' bawled out the Officer as the two men landed in the swaying boat. The next minute a wave came and lifted the boat, crashing it against the side. Loud shouts of 'hold on', but the next wave smashed it to pieces, the two occupants holding on to the boat's falls and clinging to these; we hauled them aboard, falling exhausted on the deck, one recovering but someone pronouncing the other dead. It was during this time that 'abandon ship – every man for himself' was piped. As soon as I could I dashed off to my station on the lee side, which I had left all ready to jump in, but alas when I arrived there the float had gone without me and all others on that side had pushed off from the ship out into that lonely black sea. What was there for me to do but to go back to the other side, it being further out of the water, where a number of officers and ratings remained, maybe left behind like myself. It was there I heard the engineer officer make a report to the Captain.

'The ship won't last another half an hour, Sir, there's just a thin bulkhead holding it up.'

'Very good,' answered the Captain.

I turned to a rating standing near. 'Come with me. I know where there is some timber. Help me make a raft,' and I turned, making my way, followed close by my companion. I undid a small hatch on the upper deck, crawled in and handed out several planks of timber. With one or two small lines and a larger rope between us, we ran with these on to the quarter deck, lashing each timber with the lines at both ends, allowing a few inches between timbers, also across the centre, with a bowline tying the larger rope to it and securing the end of the rope to the guardrail, slung it over the stern and watched it floating in the waves beneath us, and I must say it looked uninviting. We must have worked fast, giving all our attention to our job and not knowing what others were doing. We seemed to be working alone on the stern although many others were close by, but the job complete, we made our plans to wait until the ship broke up or made her final plunge, then dash to the stern, slide down the rope, cutting our raft adrift. We would have to paddle somehow with our hands to get it clear so as not be taken down with the ship. We then went and joined up with the others, milling around in the darkness, settling ourselves down by the sheltered side of our sloping gun shield to await events. I wondered if the thirty minutes was up, not aware of time or how long it had taken to make our raft. It then suddenly came to me there was a ship's lifebuoy in my old motor boat which was there as part of the boat equipment to be thrown out if anyone fell overboard. I quickly arose and made my way over the debris amidships, carefully picking my way to my beloved old motor boat. There it was, still covered over and lashed down. It passed through my mind to wonder why no attempt had been made to hoist it out. Perhaps power had been cut off when the engine room had been hit. The ship had stopped immediately the explosion occurred. I untied the cover, pulling it well back, scrambled in, looking for the buoy. There it was in the bows. I grabbed hold, then paused, for there plainly to behold was my large skip of fruit, looking so lovely in the starlight. Oranges, bananas, grapes and lemons. Although I was in a desperate hurry I had to stop, thinking with a sigh of regret those at home wouldn't get all this lovely fruit, then jumping out of the boat, stopping to lash up the cover again, why, I don't know. Whether my responsibilities as coxswain, my love for the old boat or in secret memory of my basket of fruit, but I replaced the cover then hastily returned to my sheltered place, starboard side, aft, where the remaining ship's company assembled. Maybe it was the highest part out of the water or the more sheltered, but I sat there now, clutching my lifebuoy as the black clouds passed over, hiding the stars and bringing gusts of wind and sweeping rain, which added to our discomfort, just waiting, looking into the gloom, while others milled around. It was then that one of the two figures walking past tumbled over my feet, picking himself up, he turned to apologise, saying 'Sorry'. 'It's all right Sir,' I replied, for I recognised the voice of the Captain, and he continued his walk up and down. I also sniffed, for someone's breath had left a strong whiff of whisky. A few minutes afterwards, further along, a group of figures seemed to be active and excitable and began to sing 'Roll out the Barrel'. I realised they had broken into the rum store and had fetched up a barrel of rum and were partaking of its contents, with the Captain having to order them to stop the noise. I weighed up the situation sitting there and decided while there was hope I wasn't going to get drunk on rum and didn't join in with some of the others who seemed to be freely indulging, so I remained just sitting and waiting, wrapped in my oilskin, with lifejacket, lifebelt and clutching my rescued lifebuoy, with the raft tied to the stern. Satan wasn't going to get me without a fight. It was then someone called out for the Captain, and I heard him suggest that a message should be flashed to a corvette to come alongside and get some of the men off. 'Is there anyone here with a torch who can flash morse?'

'The Yeoman of Signals is right here Sir,' one or two voices exclaimed, dropping back and enabling the Yeoman to come forward.

'Make to corvette if possible to come alongside to take off men.'

'Aye, Aye, Sir,' answered the Yeoman, and with his torch held high flashed the message into what looked like an empty black space. Springing to my feet, keeping close by, I watched the dot and dashing of the flickering torch. Long minutes passed by. The Yeoman continued flashing his morse. I hoped the U-boats had gone. Seeing the flashes they might send another torpedo crashing into us to finish the ship off, but this was a risk we must take as with searching eyes I looked out into the blackness of the night. Long, long minutes. Then I observed flashes of dots and dashes of an answering light, hearing the Yeoman sing out 'Corvette replying Sir, will attempt to come alongside.' I was alert, the slightest chance of rescue I was going to be ready and climbing over the ship's side, standing with one hand grasping the top guard rail, my other arm clutching my ship's lifebuoy, leaning out, ready to jump if the corvette came near enough, staring intently out into the gloomy darkness. I waited and watched. A dark shape loomed up. The corvette was making it, but alas, it shot by me some twenty yards out. My hope sank within. I watched it strike the *Springbank* for'ard with some force, noticing the white foam of the churning propellor, my watching eyes conveying to me that it had crashed into the ship's side and stopped its engines and was drifting down the side towards me, the waves lifting it up and dropping it down, coming fast down the side. In a few seconds it was beneath me. I saw a small clear space on its deck, and from the mighty heights of the tilted side of the *Springbank*, leaped for my life for that spot, falling with a wallop, but the lifebuoy I clutched took the full impact. The fall shook me as I felt a stinging pain shoot up my leg from my numbed foot. I said 'thank God', as a deck hand rolled me out of the way. I heard other's shouts and screams and above it all I heard the last scream of one who fell between the corvette and the ship being crushed to death, and I recognised that terrifying shriek, for it has remained in my ears ever since, as the death cry of my young married chum who had tramped with me up and down the deck, side by side so often. Many were striking objects in their fall, for there were only a few seconds to make one's mind up to jump, and the upper deck of a corvette is cluttered; reels, chains, hatchways, all crammed into a small area. I scrambled through a hatchway down the ladder, and sat on the lower deck nursing my ankle as my foot felt stung and useless at the moment. Others like me came down that hatchway or were brought down and laid out. One had broken his back and died the next day. Rubbing my foot and ankle I soon got going and was able to assist.

One had a bottle of rum and offered it to me. I took a good long drink, which didn't seem to have any effect, but perhaps it may have done for I don't seem to remember any more until the morning. Waking up to find myself alive, being rolled and tossed about by the unsteady movement of the corvette. I scrambled up the ladder on to the upper deck into the brightest of sunlit mornings I had ever known and there was a bit of the old *Springbank*'s bows jutting out of the sea, with the crew of the corvette bringing their small 12-pounder gun into action to sink it. Oh, it was a good feeling to find myself alive and well after the ordeal of the night.

The remainder of what was left of the convoy were well ahead and out of sight. Then hearing a drone I looked up and there was the old Focke-Wulf flying round to have a look, so close, so clear, its silver body gleaming in the bright sunlight. If only it had flown close like that with the *Springbank*'s 4″ guns! We could have brought it down easily, but now it came so near you could almost hear the plane's crew laughing at our plight, exhilarated by their success. The corvette brought its gun into action, but there was no danger to the German plane. The burst from the shells was as far behind the plane as the target was to the ship and I felt a terrible urge to pull away the trainer of the gun, jump in his place, assured that I could do a lot better, but after a last look by the German plane, seeing the last of the *Springbank* just bobbing out of the water now, it flew off, no doubt to catch up with what was left of the convoy. The corvette, as soon as the wreck dipped under, put on speed, leaving behind that part of the sea, with its graveyard of ships. I went below to have a look at some of my mates and believe me my eyes opened wide, for I ran into an angel. Had I passed over from an earthly life? I stood back in surprise and wished her good morning, for this was a lady adapting herself to the conditions and doing her share of work of cleaning up and attending to the injured. I took it she was a survivor like us. I never knew her story as I attended then to one of my chums who seemed badly hurt and insisted I take the two bottles of rum he had with him. I put them in the pocket of my coat hanging it up in the flat along with the others. This flat where we stretched out and bided our time, making ourselves comfortable, for we didn't want to interfere or get in the way of the corvette's crew who had saved our lives. Later on when I thought I would celebrate with a swig of rum from my two bottles, I discovered they had both gone. 'Most disappointing' I told my chums who were going to share them with me, but worse things than that happened at sea!

We were meeting it a bit rough as the corvette was going flat out to catch up, but it wasn't until the second day that we came in view of the convoy and were thankful to see that there were still some of the merchant ships left, so well guarded now that Jerry wouldn't stand a chance, not only ringed with naval vessels but overhead flew numerous escorting two-engined planes, for we were nearing the Irish Sea. That night I lay down in the flat of the corvette, relieved, happy and thankful, feeling safe on this bouncing flat-bottomed corvette trying to keep station on its unsteady course over the ruffled waves, getting nearer and nearer home.

In the morning a number of us were landed at an isolated pier off Milford Haven, climbing up many steps to reach the cliff top to some huts, and after being given a meal, put on a train to Chatham, arriving early next morning in London, having to walk across London in seaboots, wearing the same old clothes with our week's growth of beard, morning road sweepers and porters asking us if we had been shipwrecked but remaining dumb to some of their questions, entering the main gate of Chatham Barracks and for once not saluting an officer as without hats, and in seaboots and old coats, we presented ourselves at the office window as survivors, the Master at Arms for once dropped his commanding voice, and after taking particulars, we spent the morning going round the different officers and finishing up with being issued with a sprog suit again and sent on 14 days' survivors' leave.

Now I walked through Gillingham to my home. I was a sprog again. My sailor suit fitted where it touched, with those narrow legs – only sixteen inches, when the sweeping bell bottoms made by the civvy outfitters were twice that size. But my ordeal was over. My fourteen days' leave. What a surprise to meet the wife! She would be at work at her market-gardening nursery which I had to pass on the way home. I must call in and enquire her whereabouts.

'Sent to one of the greenhouses,' I was told, and soon, as I entered, beheld my loved one, working away with her sister, a ribbon tied round that shoulder-length hair to keep it out of those hazel eyes, as she bent down attending to the growing plants, unaware that her sailor husband was creeping up on her.

'Good morning,' I exclaimed suddenly.

Heads looked up.

'What are you doing here?', she asked me and I noticed a catch in her words.

'I have come to take you on holiday', I hastily replied, 'for a whole fourteen days, starting right now.'

'Why are you in that sprog suit?' asked her sister.

'Oh, our old ship went down and I just stepped on another which brought me home and now we have been given some leave to await another and I didn't have time to change.'

'Well, you go and sew some badges on that suit and look a respectable sailor again.'

'So I will, but I am taking my wife with me,' taking my wife's hand. We walked home.

'I am a survivor,' I said when out of earshot of the others.

'I know,' she answered quietly.

'How did you know?', I asked in surprise. 'I only arrived in barracks and fitted up and straight away came to you. How could you know?'

'A wife's intuition. Mr Clarke who deals with the service mail in the barracks quite suddenly one morning on seeing me jumped off his bike and asked me if I had heard from you. That was all that was said but from that moment I knew you were in some dreadful trouble and have been fearing every moment.'

I gripped her hand more firmly, for we were going through the wooden gate of the back garden, where many nights a few years ago we had stood with a lingering kiss with all the promise of a golden life together. Now Isobel was in her working clothes, I in my sprog suit working and fighting to keep ourselves and our country intact. It wasn't until we were together in our room with our small son that we could free ourselves of the tension of those last few days and lie relaxed, slowly realising we had come through another ordeal and were still together.

CHAPTER TWENTY-NINE

I stood on the upper deck of H.M.S. *Gambia*, a cruiser of the Colony class, to which I had been drafted. It was a modern cruiser of 8000 tons displacement, with a speed of over 30 knots, its main armament being four triple turrets of 6" guns, two 4" dual-purpose twin guns on the port and starboard sides of the boat deck, and numerous smaller weapons of close range spaced around its upper decks, with triple port and starboard torpedo tubes. A formidable fighting ship, with tremendous firepower.

We had commissioned the *Gambia* at Liverpool, done our work up at Scapa Flow with the fleet, and now it was a fully trained fighting unit, ready to serve anywhere. This was early in 1942, when Britain was threatened in every part of the globe. The see-saw battles of our Army in the Middle East, the Navy's role being to keep Malta supplied. The whole Mediterranean was a fighting lake, the U-boats were taking their toll of our Atlantic convoys, German troops were battering at the doors of Moscow, and now the Japanese had captured Hong Kong and even the fortress of Singapore had fallen. A link-up of the German and Japanese forces somewhere in India was a fearful possibility.

What an enormous, amazing ambition this little strutting Corporal Hitler had, now Chancellor of Germany, Commander-in-Chief of its forces, about to conquer the world. There only remained that final swoop. Russia was almost finished, America had been caught napping in her isolation by the Japanese at Pearl Harbor and Britain was almost hemmed in and starving. Two major German warships had even made a dash and got through the English Channel.

Now we were entrusted with the mighty task of escorting home the Australians, who had fought so well in the Middle East. But now Australia was threatened by the Japanese, so naturally the

Australians wanted to defend their own homeland. To see our magnificent liners the *Queen Elizabeth* and *Queen Mary*, who dwarfed even the largest battleship, boldly steaming through the waves, their upper decks lined with troops! What a prize for a waiting U-boat; indeed, worthwhile risking a pocket battleship to finish off our Empire troops and to sink the two great liners, which were not only the pride of the shipyard workers of Britain, but dear to the whole nation. It would be a disastrous blow if this did happen; these two liners' speed alone made them a match for many warships, but many other large ships were in this convoy, so no doubt their speed would have to be cut down.

We steamed across the Atlantic, down the coast of West Africa, round the Cape, where the weather turned foul and ships were no longer visible through the drenching rain and spray.

Two ships were damaged by storm, mine or torpedo. All I knew about it was that we had to stand by until the sea calmed down and smaller vessels arrived from the Cape to tow the damaged ships safely into the harbour. They may not even have been in the convoy. As soon as we were released of our duty, away we sped to Durban to be united again with the troopships and restock and refuel and proceed on our final run across the Indian Ocean. Days of blue sea and blazing sun with the convoy now cut down and steaming faster. I could see from the *Gambia* the upper decks of the two *Queens* thick with figures swarming over every open part, gazing across the clear ocean and drinking in the fresh sea air. Only a few weeks before they had advanced across the yellow sandy desert of Egypt. Now they were going home. Care and fear were just under the surface, even when three-quarters of the voyage was over. The last few days and then Australia. Goodbye to the two *Queens* as they steamed down the coast to Sydney amd the *Gambia* put in at Fremantle, the port of Perth, where I felt a thrill to set foot on Australia, this country which had given a new home and fresh start to many of my countrymen of the same class as myself. Many had no choice and as convicts had been transported to this great land. Yet those men, without hesitation, had volunteered to lay down their lives when their mother country was in peril.

Having been badly affected by Australian alcohol on my first trip ashore, I was relieved to find an empty hall attached to a church, a sheltered haven for an unwell sailor. On a settee I stretched out to become dead to the world, and at length a gentle whisper reached my ears as if the voices of angels were safely speaking in this holy place. I lay for some time, then gradually opened an eyelid to observe a homely gathering of motherly ladies.

'Don't speak loud or make a noise or we shall wake him up. Let the poor sailor sleep on. What a clear and fresh complexion he has and so young, poor fellow.' I heard the words of praise from the hushed voices of the motherly ladies as I pretended to sleep, but even this pretence made me quail within. When I woke up I was made welcome, and after being shown where to wash and refresh myself I sat down to tea with these kind ladies, who pressed heaps of tasty cakes upon me with homely cups of tea to drink, with them wanting to know what England was like.

After I had given a description of my native countryside, the ladies fell to talking among themselves, picturing my England, thanking me, and offering me an invitation to their church hall dance. I went off to the ship with a satisfied feeling; even if I had made a bad start on my first run ashore in Australia, I had more than made amends.

I attended the dance given at the church hall and was enchanted by the old-fashioned style of their entertainment, as if the courtliness of old England was still maintained by its English descendants. This part of the world had not been affected by modern dances such as the Charleston and others. In gentlemanly fashion you graciously asked the favour of the damsel to honour you. If accepted, you ceremoniously escorted her, politely taking your time to join in the slow waltz, holding her at arm's length as you whirled around, then gallantly escorted your blushing partner to the seat by her waiting mother, and if favoured by an inviting smile you pressed your attentions further, joining in the family group, maybe leading to an invitation to a respectable home.

After a few day's stay, we sailed out of Fremantle, leaving behind an unspoilt Australia; also four ratings who maybe had genuine excuses for having missed their ship, having had an accident or lost their heart. What would happen to them? A firing squad for an act of cowardice or let off for not knowing the responsibility of war? It would be for the higher authorities to pass judgement. Four more men with family ties at home would have to replace them. I had no time for deserters in war.

Months of sailing alone in the vast Indian Ocean, protecting shipping from the advancing Japanese. We put in at Bombay, Colombo, Durban, the Cape, Mauritius, Mombasa and Diego Suarez in Madagascar. We docked at Durban, the South Africans making us welcome, but in the Afrikaners of Dutch descent still smouldered a hatred, which they handed down to their children because of British treatment of them in the Boer War of 1900–1902. They were part of the white minority who govern states of Africa, keeping the black Africans as workers; I feel for their safety and wonder for how many centuries a few whites could remain masters over a large number of black people.

Mombasa: where we were limited to a few hours' shore leave, and went out to sea one day as escort to one of our aircraft carriers, training her pilots in taking off and landing; to my horror there were eight accidents during that afternoon's trials, resulting in plane losses with pilots having to be rescued from the sea. If these accidents happened when the sea was of millpond smoothness, what took place in actual combat when the sea was rough?

Colombo, with its warm blazing sun, the fireflies dancing in the night air. Bombay, the gateway to India. However, before entering Diego Suarez's natural harbour, we learned that it had been captured by a combined force, with the destroyer *Anthony* steaming through its narrow entrance to land marines on its wooden wharf before this French-held seaport actually fell.

It was from there, one day, that we sailed out, proceeding down the East coast to meet up with a large force; we were given to understand further military operations were necessary against the island, as the French were unwilling to co-operate with us, although we had taken Diego Suarez, so a combined force was sent to occupy Madagascar and remove the threat to South Africa.

The large force surprised me. We had been alone so often that I thought we were the only seagoing warship guarding the whole Indian Ocean, even in this remote part of the world. It gave us a little flicker of hope. Now well into the third year of war, with all the defeats and setbacks, we were gathering strength, so that I remembered the words of old Stripey, a three-badge sailor who had told me to stop worrying about the results, saying Hitler, who had failed to take England at the beginning, was just blowing himself out – and to see so many of our ships certainly gave me heart. Which soon sank again! Tamatave, the Island's chief Eastern port, was our objective. We sailed right in and the *Gambia* dropped anchor only a few cable lengths off the shore, and as our ship swung round, riding at anchor, I could see clearly through binoculars, soldiers crouching in trenches. How could our Admiral present us as a sitting target? It didn't help my fears when a motor boat with white pennant flying approached the beach, only to hear the rattle of a machine gun open up with a hail of bullets spattering the water round the boat, which made a swift hard over and returned to the flagship. Then over the loudspeaker came the information that the town was being given twenty minutes to surrender. 'Crumbs,' I said to my sight setter, 'I wonder how many guns they have got trained on us?' Time seemed to pass so slowly. Five – ten – fifteen minutes. Still nothing happened. I watched the soldiers waiting in the trenches until from our communicating number came the order 'Stand by'. With guns at the ready and every nerve tingling, there was a countdown of the last twenty seconds: nineteen, eighteen, seventeen, sixteen – breathlessly – nine, eight, seven, six, five, four, three, two, one – Fire! Flame flashed from the triple barrels, followed instantly by the cordite smoke as the thunderous, deafening roar swept across the bay, the clatter of the gun breeches opening, the ramming home of the next round, as I braced myself, bringing my pointers in line and steady after the jolt of the first broadside. But what was that? The ceasefire bell was rapidly ringing, before we could get the second round off. 'Open interceptor,' I heard ordered from the captain of the turret. I took a swift look through my binoculars and shouted to those in the turret 'White flags are going up everywhere. The whole town has surrendered.' Planes from the aircraft carriers roared overhead and jubilant voices could now be heard in our turret, when only a few moments before your own thumping heart had sounded so loud.

We left the coastal town of Tamatave, now that it had surrendered, with the army, I suppose, taking over and proceeding inland, the *Gambia* to continue her patrolling rounds, visiting Durban for docking, when our engines needed overhauling, and the long, weary months slowly passed by. Occasionally we picked up mail from home and perhaps we were fortunate in a way in not encountering more action, but the time seemed to hang heavily on our dreary, weary seagoing trips. Only in Durban was there any sort of life.

Slowly the tide of ill news reached its highest level, and we heard news that our enemy was meeting stiffer resistance. The

Russians had launched an offensive at the eleventh hour, our Eighth Army were driving the Germans from Africa, the American naval planes were sinking the Japanese fleet. Our hopes slowly rose and the rumour grew stronger on the mess deck of His Majesty's cruiser *Gambia* that we should soon be sailing home. Had the mighty gallant Navy reached its zenith of power; were the speed and range of the plane now playing the master role in this World War? For the news relayed to us was that the R.A.F. were massing large bomber raids on German industrial towns. The threat of German and Japanese meeting up in India was fading, the cruiser *Gambia* was being recalled and its ship's company rejoiced as she began her voyage home, receiving the gift of two tons of sugar from the small island of Mauritius, and putting in at Bathurst, the seaport and capital of our oldest colony Gambia, where I accepted a trip up the River Gambia on a steamer. We went several miles until it anchored in midstream to allow us to wade ashore over swampy marshland into the jungle, where we came across a native village of thatched mud huts. We went further into the bush, to come upon a few banana trees. There some of us with our knives hacked away at a cluster of green bananas; the wood was tough and green, but the blade of my knife after half an hour's sawing cut through the stalk and I came away with a large bunch, only to be stopped by the native village copper, with a man whom I took to be the owner of the plantation. The latter protested profusely, but as there were a hundred of us to two of them it was a one-sided affair; I gave him ten shillings and others likewise, and I am sure he received more than they were worth.

Back on board I nailed the bananas up in a box, deposited them in the cable locker, and when we arrived at Plymouth some ten days later I despatched my box by train, so that they arrived home just ripe for eating, a rarity – for none were shipped into the country during the war years. Two days at Plymouth and then we sailed to Liverpool for docking; I sent a telegram to Isobel to come and join me.

In Liverpool, despite all the bombing, the busy life of a large seaport carried on, and it was here amidst the turmoil on Lime Street Station that I met my Isobel. We took a little bedsitter in the middle of the district of Wavertree, an area of many little streets and a thriving local, where chat over a glass of beer soon restored my depressed spirits. There was our little David, now four years old, trotting along and talking. With these few episodes of family life, it made my heart bleed to think that war was still going on, but I realised how fortunate we were, for there were thousands worse off than ourselves, many who had lost their homes. Walking across the bare patches of land with little David trotting in between us, I thought of the families who had lived in the houses that were now empty patches. Had they all been wiped out? What would have happened to me if I had returned to find a bomb crater where our home had been; many of the men in the Forces must have had that experience.

When our little son was fast asleep in the small makeshift cot by our bedside and we lay in bed, my arm around my beloved, I spoke of all the passing fears that had swept over me, when I had thought of home.

'I never want to let you go,' I told my wife.

'Get away! What about those Australian girls who you say were so unspoilt, with all the promise of flowering loveliness like fresh unopened rosebuds?'

'You must understand', I replied, 'that it was the one place in twelve months where I found real and true life existed, after seeing the poverty of the tropical countries. There will be so much to put right when the war is over. But what will those countries think of us when the war is over?'

'Please my love,' Isobel carried on, 'all those long months I felt so lonely my heart ached, so forget about the past and the future. Kiss me and let us live for the moment and the few remaining weeks we have together.'

CHAPTER THIRTY

What was happening in my peaceful old village where the wife, son and myself had come to relax on my fourteen days' leave given by the Admiralty to those who had done foreign service? There was anxious concern felt by the village community for the hungry time the country was going through, now fully aware that the ration of food must have been stretched to the limits if the village felt the alarm and threat of starvation. The people needed some encouragement. Churchill's speeches of September 1943: 'Now those who sowed the wind are reaping the whirlwind. There is no halting place – we must go on,' had given us courage and determination to fight on. The war effort had brought people together. The poorest villager who had set his garden with vegetables to help him through the hard winter months, when production was at its lowest ebb with work and money scarce, now found himself sought after by his country cousin, who offered to pay a good price for any extra eggs, fruit or vegetables, while the porker kept in the pigsty at the bottom of the garden was a goldmine to the poor labouring man. Even the village middle class, who lived in brick houses set in their own grounds, rubbed shoulders in the local pubs with those who lived in old thatched cottages and small terraced houses, who drank their cold four-penny pint, who fumbled around to pull out his twopenny packet of Woodbines or lit his pipe from the taproom fire to spare a match, was now asked how his son was getting on in the forces, which often led to a more friendly atmosphere, and a small transaction of a few eggs, hare or rabbit to stock up the middle-class pantry.

The poor villager now found he could make a few more shillings much more easily, to enrich his humble existence. At last his life of hard labour was having to be recognised and rewarded, for it was rumoured in the village that often the butcher's van or cart carried back more meat than it brought, being supplemented by the few men who had survived the call-up by their art and skill in catching the squire's game.

But drawing them more together than anything was the Home Guard, which was formed to defend Britain against a German invasion. It brought together the squire, the commanding officer with his military bearing, and working men, left behind because of some physical defect or because of special skill. They were thus able to do their bit to teach or be taught by those old soldiers who now became sergeants and corporals. The few men and youths left behind in the village were taught to shoulder arms and drill, and this led to their skill on the firing range, the real test of each individual's ability. From as early as they could remember they had thrown conkers and stones and had used catapults, which had led on to early experience of firing a gun, their natural country skill enabling them to creep up silently and patiently, not affording to miss many times as they pottered the rabbit or brought down the pigeon. Now the ·303 rifle gave them a military bearing, a pride of place in our country's fight against its enemy. In the parks under high trees, camouflaged rounded huts now stood, filled with soldiers being prepared for that great day when we would launch our main offensive across the Channel, but these now added to the village life bringing in a new and enlightening friendship.

My wife and I did some digging and weeding and generally helped my father, and in the evening we played cards or darts, and lingered over a couple of pints as we discussed the latest news. Now one special pleasure of homecoming was the rare extra rations my family had been able to hoard up for such an occasion.

One night my father said: 'I will give a little celebration tonight. I have a joint of pork saved from the last pig killed, and after I have called time I will ask a few to remain behind.' So, just before he called 'Time gentlemen please' there would be whispered instructions to certain ladies to slide into the kitchen. Soon Doris, May and others would assist Isobel and my sister, Nellie, heaping up large dishes with pork sandwiches. In would come many of the long-standing customers of the King's Head, even to the village constable, who cautiously stepped in at the back door. My father, having tapped the fresh barrel, filled the glasses of all those who had squeezed into our kitchen.

My leave at an end, I walked into Chatham Barracks; for how long, I wondered? I could put my name down for higher gunnery rate. That should keep me in a few weeks from having my name going up on the main drafting board in the drill shed, and so with that little bit of luck, I found myself on course for a Layer Rating Second Class 'L.R.2', to be taught more about the training and laying sights of the big guns. A native again, on night leave three nights out of four, three weekends and then a duty, and not much fear of an air raid, so the next few weeks saw me strolling home, getting accustomed to living like an ordinary married man, setting the alarm clock to wake, joining the hurried footsteps amongst the many dockyard workers and sailors.

Again, the mustering and classes, the rigid discipline still maintained by the Gunnery School. Occasionally the sirens wailed, for raiders still sometimes flew over to remind us of the power of the Luftwaffe.

I was taught the laying and training system, the finer gyro sights of the main director, testing our aim and judgement in hitting a moving target under severe weather conditions. I was all right at the practical, as our sights rolled through on the moving mechanism, but when it came to explaining the system, how it all worked, I found the complications never worked out clearly in my head, but anyway, the passing-out officer awarded me enough marks and I took home a badge with two crossed gun barrels and a star above and below to be worn on my left sleeve, for on my right sleeve were the anchor of a leading seaman and two good conduct badges. Anyway, that should make a few look up to me, if it was only the young sprogs just joining up, I told the wife as she put the finishing stitches round the badge, standing back to see it wasn't tilted, as a sailor's wife should.

'You know what this will mean: a ship,' then adding, 'and I'm just losing my rolling gait.'

'Well, that's what you joined for. To sail the seas; and that is what I will have to put up with,' was Isobel's comment, and we would leave off talking of ships, as it was too touchy and sad a subject to dwell on.

About this time we received a letter from my sister saying my father had been allocated a house in the village and would be moving out of the King's Head. This gave me quite a jolt, with many thoughts going through my head, turning them over in my mind, and talking it over with my wife, for I saw the King's Head as some sort of fulfilment of my young ambition, when I declared to my aunt that I would have a farm and a hundred pounds by the time I was 21; for this village inn had all the buildings, a proper wall in the cattle yard and stables, a garden and field at the back, with a meadow too by its side, which my father hired. Now I weighed all this up and sorted things out. Lucky for me I was home because the more I thought of it, the stronger grew the longing that I wanted to make it our home, and I could hardly wait for weekend leave. When I reached Wilbraham to consult my father and others of the family it was agreed that we should make our home at the King's Head, with my wife looking after my father. We managed to move our furniture with a hired lorry from the village during Christmas leave.

Now I was back and settled in the village with my wife and child; I was now a Leading Seaman of the Royal Navy, who had travelled to many foreign lands and was now in the middle of the greatest war the world had ever known. Would it come to a successful end so that I could return and start to build up life again in my boyhood surroundings? I had a feeling that somehow I was laying a foundation for when the war ended and we could begin again, and when I went to the village church that Christmas morning, in my silent prayer I asked my God that this might be fulfilled.

CHAPTER THIRTY-ONE

At Plymouth I joined H.M.S. *Melbreak*, one of the small destroyers named after a famous pack of foxhounds. Now these Hunt class destroyers were turned out to do just that, only their hunting was in the English Channel where lurked the German E-boats, speeding out into the Channel and lying in wait, torpedoing anything that came in sight.

Within hours of leaving the comfortable, swaying seat of the train, I was sitting in my director's training seat, which my gunnery course had trained me for, head buried on the foam rubber protection piece of my powerful binoculars, sweeping over the restless waves. There I would remain all night, alert and ready. No longer was I on the defence, but in the forward line, ready to pounce, driving the enemy back into their harbour.

We did a regular sweep of the English Channel, keeping the sea clear for our coastal merchantmen, giving the enemy no time to lay his mines or lie in wait.

It was only when a defective engine needed repair or boiler cleaning came round that the ship's company was able to get any leave. However, with the ship always at short notice boiler cleaning came round quite often, and away we would be off to our homes, united with our families. As I walked the last mile or so I was surprised to see the Home Guard turned out on duty at every footpath and by-road. 'Why have you turned out? Have I a guard of honour?', I asked Dick and Willie Wilson, two local men I had known from boyhood. 'We have been called out to standby for a possible enemy paratroop drop,' they informed me. I offered them a duty-free cigarette, thinking to myself that we had just come from sweeping the Channel; if there had been the slightest danger of any attack on the part of Jerry, no boiler cleaning leave would have been given.

During the few months I was on the *Melbreak* we never had any actual ship-to-ship engagement, but there were some exciting and breathtaking moments. When sailing along in the darkest hours of the night we felt a rise in speed, the *Melbreak* now leaping along as the order came quickly over 'Stand by to open up', followed rapidly by 'prepare to ram'; those not secure at their quarters hung on to the nearest object, waiting for that momentous second of impact. A large 'ping' was picked up on the Asdic. All set – about to sacrifice lives and ship in honour and glory as the reported object grew nearer and larger. Then suddenly the shouted order 'hard over', throwing the *Melbreak* on its beam end and causing damage and a hard knock to ship and lives, leaving us wondering why. Then the rumour spread around on the mess-deck the next morning: owing to a navigational error we had almost rammed Eddystone Lighthouse! Or the occasion when on our own, giving air protection to coastal shipping, a large aircraft approached showing no friendly signal. Immediately the director swung round, picking it up in our binoculars. Having learned that there was no reply to our signals, the Captain ordered 'Open fire!'

The *Melbreak* was just about to get its first kill. We would get it the next round. The tense long moments of waiting for the fire gong to press that trigger but not the ding, ding. It was the rapid ringing of the ceasefire bell, with the operator hurriedly singing out 'Friendly signal showing'. We sat back and watched one of our own planes wobbling its wings frantically, indicating its friendliness. No doubt there would be an enquiry, somebody would get blamed for being asleep, but we could take no chances with a plane flying overhead.

The incident when I failed in my duty to ship and country, being detailed to go into the gunnery school in Plymouth Barracks with the gunner's mate and another for a refresher run on the gyro sights. This was duly carried out. It had been a nice little trip ashore, but what must we do when leaving barracks but wander off to a canteen on the cliff top overlooking the Hoe, from where we could see our flotilla tied up to the wall. How nice it was to down a few jugs of sparkling ale, with the beautiful scenery of the island harbour to gaze on. Even Drake couldn't have been more composed over this bowls than the three of us over our ale, when our mate gripped our shoulder and pointed out to us that the flotilla was getting up steam; the ships were preparing to go out! Some information must have come through, because when we left no sailing was anticipated and engine room watch had been cut to four hours' notice, but now the engines were being stoked quickly, if the dense smoke from the funnels was anything to go by.

'Ah well, we will have another pint each,' said our gunner's mate. 'Drake finished his game of bowls so we will finish our pint,' and so we took our time, although I felt a bit uneasy as the smoke became thicker, thankful when the P.O. got up to go back, and we only just made it, the gangway being delayed for us to scramble over. Then the flotilla was under way, out through the harbour as darkness fell.

But now I regretted my pints of ale. My strength deserted me; I couldn't turn my training handle, and I asked to be relieved as I was feeling so unwell. I crawled along to the sick bay attendant who took one look at me and said 'Take a look at yourself in this mirror' as he quickly mixed me a glass of fizzy white mixture which I drank. Then I sat on deck, fighting for air. But worst of all, when I had sufficient strength restored and returned to my post, the officer would not let me in my director. 'I am all right now,' I repeated, but had to return to sit on the upper deck. At about 0430 we swung away and flat-out belted home to the sheltered haven of dear old Plymouth.

Many times our flotilla had put to sea in the evening light, and on our return in the grey light of the morning we had always seen a cruiser tied to a buoy and would mutter 'it's always the little ships that do the work', but one evening our pack of foxhounds streaked out to hunt, taking the cruiser with them. The *Melbreak* failed to make it owing to an engine defect, but in the morning as we lined the guard rail to watch them return, alas, the light cruiser was missing.

I was recalled to Chatham, thankful in a way to be safe from this dangerous area, but regretful in another, as there is a tingle of excitement at being near to death, especially when the life of your country is at stake, and all its people are looking up to you.

I was now back in barracks, which meant weekends of home life, and for the first time in five years there was a general feeling of being satisfied with all our country's efforts. The Japanese were being driven back in Burma, the Allied Forces were fighting their way up the Italian mainland and there was terrific bombing of occupied Europe. America with its great resources, the key to future world power, had been forced into the war after their fleet had been sunk in Pearl Harbor. Now their troops were widespread, many being embarked to Britain, their air force taking over whole airfields; England was becoming a fortress filled with Allied troops, especially in East Anglia, my native district, which was dotted by United States Airforce stations. Their personnel found our English inns were friendly meeting places, and soon they were welcomed into homes and families.

By this time I had become a Petty Officer, wearing crossed anchors with a crown above, and often wondered how I had reached that position, for the giving of orders was not a strong point with me. In my examination subjects I only had pass marks, but perhaps because I persevered in my duties I was considered trustworthy and reliable, which made up for my lack of brilliance. To further my career, I requested to go on a higher gunnery course for an L.R.1 (Layer Rating First Class), which would enable me to stay in barracks a few weeks longer.

During this time, in between the weekend leave I would walk home to my mother-in-law's home in Gillingham on night leave. We were company for each other, and I had a nice comfortable bed to sleep in. By now the throb of a night raider's engine was hardly ever heard, and one could practically be assured of a night's rest. But what was that drone of many aircraft approaching – going over – waking me up? No fear of alarm as I lay listening to them; by sound alone one could tell that these were friendly planes. But now, continual waves of planes were flying over, arousing neighbours to rise, open their windows and even to go out on the pavement at this early hour. I muttered 'Why do they have to practise at such an early hour when I do get a few days to stay in my home depot?' Now my neighbours' voices carried across the street. This must be D-Day. Then I heard 'D-Day' repeated. I warmed with delight and lay letting it sink in. At last! So I lay thinking on that revealing morning of 6 June 1944, as I listened to the now pleasant drone of the waves of planes going over. Then a sudden alarm. From the drone of the continuous waves came the terrifying whine of a plunging aircraft, seemingly right overhead. Then I heard the thump as it hit the earth. I rushed to the window to see dense black smoke from the rising ground close by, and immediately jumped into bed again, pulling the blankets over my head to avoid glass splinters from the windows as the bombs blasted, having read that this was the correct thing to do, for the shattering rumble of the bombs followed. No doubt the blast was muffled by the crater made by the falling bomber, as only the window frames shook, but the incident had a sobering effect in bringing home to us the supreme sacrifice which was being made that day to restore freedom to those countries who had been overrun. That morning I arrived in barracks, joining in the glowing, excited feeling that at last Britain was putting her whole weight behind the final assault. And as the official announcement was declared on that memorable day we waited with bated breath, fearing yet eager, carrying on our everyday duties, praying and hoping for success, yet knowing thousands of our young manhood would never return.

And as I continued my higher gunnery course, went home to my mother-in-law's in Gillingham on night leave, down the line to Wilbraham at weekends, the bridgehead in Normandy was built up. The many troops stationed around the countryside rapidly dwindled as they were ferried over to France, the hopes and prayers of all their English friends with them. Much of the life went out with the troops gone, and the village settled down to its more sober way of life.

Could we now take a rest from the fear and agony of the bombing in the south-east corner of our land now that our troops had successfully crossed the Channel? Freedom at last from the sky? But this was not to be. Berlin claimed it was the beginning of the day of vengeance and filled the sky with flying bombs. To see these flying through the air, carrying a warhead containing over 2,000 lb of high explosive, especially when it was directly overhead, was a most frightening experience. Sometimes we were on the parade ground, marching up and down, and over would come a flying bomb chased by a fighter; we would look up, still marching in step, praying for the fighter to hold its fire until it had passed over. But worst of all was the effect at night. For miles away you could hear the noise of these doodlebugs. Clearer and louder the noise would come, as we prayed for that engine to keep going, for when that stopped then you knew the flying bomb would plunge to the ground. Long breathless minutes as the 'phut phut' sounded, then the rising thankfulness as you sensed it was level, passing over, and the feeling of relief as you looked at each other now that the noise was fading away. This could happen more than once each night. I felt that in certain areas of England the civilians suffered more than a good many troops.

But now I had a quick draft to H.M.S. *Mauritius*, a sister ship to the *Gambia*, of the Colony class of cruiser, joining it overnight at Portsmouth, and in the morning we were under way, heading for the invasion beaches of Normandy, where our allied troops were locked in battle with the German Panzer Divisions, thus slowing down our bridgehead. That morning before dropping anchor in the shelter of Mulberry Harbour, the German defences were subjected to a 1,000-bomber air attack, the sky filled by groups of our bombers going in unmolested by the German Luftwaffe, that once proud force that had boastfully ruled the sky.

Our orders were to bombard the German positions under the direction of the army forward observation post, which would send the range and direction in which to fire. My position, now I held a higher gunnery rate, was in the director, transmitting trainig down to the turrets for them to follow. We fired salvoes from one of our turrets all through the day, continually shelling and keeping this up for a week, including a return to Portsmouth for re-ammunitioning. Thousands of shells we must have pumped into the German defences at ranges from 14,000 to 20,000 yards, until finally came the breakthrough and our soldiers were at last chasing the Germans. The *Mauritius* pulled out and after refuelling and ammunitioning, took up a position off Brest in company with another cruiser and a number of destroyers, the force to give assistance to the land forces if called upon and to sweep in the vicinity for any escaping vessels that might put to sea from the French ports that were now being attacked by our land forces. Immediately the order came to shell Dieppe, a strong German-held French port, but on our way this was called off as it had fallen to our troops, and I marvelled at the speed of our forces.

After this we took up our position in the Bay of Biscay, there to patrol off the coast. We sailed in the smooth and calm sea that prevailed at that time, perhaps feeling that at last we could leave it to the army to finish off the enemy as we basked in the warm days of summer, lying around or sitting on our turret as we viewed from afar the rising land of the French coast, but to get too complacent is most dangerous, as on this pleasant morning, surrounded by the glassy sea – what was that flash from high up the coast I thought I noticed, or was I mistaken? The sun glittering on some reflecting object – just imagination, flitted through my mind. Next moment, this flash was no figment of the imagination as two splashes straddled us and figures that a second before were browsing in the sun now sprang up in rapid alarm, jumping into the turret in as swift a movement as ever one would see. The alarm bell rang out and I had my turret swinging round with the speed of the ship increasing and doing a 'hard over'. The destroyers were pouring a smoke screen, with us firing a broadside as we dashed away to put ourselves out of range of this coastal battery.

Of course, there was confusion on the lower deck. One blamed the other for being asleep. What went on amongst the gold braid of the higher ranks I do not know, but I was sent for soon after by orders of the gunnery officer and told I was to change my gunnery position from aft control to trainer in the director immediately. This was a lift up for me, to take the place of a senior L.R.1 who no doubt had to take the blame for the confusion that followed that lapse of concentration on that sun-drenched morning.

I made my way up to the bridge where all the decisions were given by the supreme commander, the Captain himself, going even above him to that conical-shaped tower which revolved round so that I could look down and around. I took my place amongst the gold-rigged officers to settle in my seat, looking through and adjusting those powerful binoculars that showed the cross wires with that perpendicular wire that I had to keep on the target, steady and accurate, so that all turrets could rely on me to give a concentrated broadside. There was the commissioned gunner who sat on my left side to elevate and press the trigger to fire all turrets simultaneously as the centre of our cross wires passed through the target at the appropriate instant, allowing for the roll and pitch with the knowledge that only comes to men who know the feel of the sea. Two more officers sat at the back and slightly above, giving firing corrections. This then was the number one fire control of the 6" cruiser that the German land forces feared and which could rival any German ship of equal size. This was the awe-inspiring position I found myself promoted to in a hurried appointment as we carried on our full operational duties with the night closing in and the sweep of the radar beams probing the darkness.

In the middle of the night 'Stand by' was ordered from the bridge and all the guns came to the ready, for something was being picked up and plotted, and an attacking position was being taken up. 'Fire when on target' came the order, as star shells fired from the destroyers came dropping out of the black sky. Under the glow of a star shell came the darkened outline of a moving vessel. Soon after, the *Mauritius* was shuddering under the impact of twelve guns. Masts, funnels and other debris flew through the air as black bursts of smoke billowed out, the first broadside devastating its target. Altogether eight ships were blown out of the water that night, and as dawn broke we steamed off, so as to be some distance away at full daylight. We viewed the wrecks of our night's work, lying on the shallow bottom of a bay. There was no activity on the beach or any movement I could detect from the numerous dwellings ashore. Being caught up in this situation was probably more dangerous for the locals than being in the forces.

That morning we withdrew out of sight of the land, slowly sailing around waiting for the night to close in, when perhaps more ships would be trying to escape from Brest and other fortified French ports, which the Germans had held since 1940 and which were now being battered into submission. After five years of having to take it, we felt at last we were giving it back.

That night, as darkness closed in, the *Mauritius* moved at speed towards that sheltered bay, her crew all waiting expectantly for the events of the night before to happen again. We did not wait in vain. The destroyers turned away. Then the calm, clear, decisive order 'Open fire when on target', for the flash of the destroyers' guns pierced the darkness, sending star shells skywards. Our eyes were pressed firmly in the binoculars, sweeping the black surface, the whole main armament at the ready and following our every movement. 'Train right,' I hear, as the light of the star shell shows the bows of a merchantman sailing into the broadside of the *Mauritius*' guns; a few seconds later the vessels flew apart. Dense smoke curled over the hull. Am I committing murder? I am thinking to myself. There are human lives aboard and we are blasting life to pieces as well as the ships. No retaliation, no chance to surrender or abandon ship. From the outer darkness they emerge into the lighted area of the falling star shell, to be blown into eternity. I couldn't help feeling a bit uncomfortable with the searchlights from the destroyers now playing on the vessels. It seemed as if they were trapped, with the destroyers positioning themselves at either end of the line, for these escaping ships to be picked off by the *Mauritius*' guns. At least they will have a chance to jump overboard and swim to the nearby coast, I was thinking, as our guns thundered out their deadly destruction of each one in turn. Finally the last ship was lying on the bottom, or what remained of her, and we turned away. Two nights of good work, for that is what it's like when you have the enemy running away. God knows, we had about five years of having to face what the Germans threw at us.

Our task completed, the *Mauritius* headed for our old wartime base, Scapa Flow, which we had first sailed into at the commencement of hostilities. There was still the North Sea, the long Norwegian coast where the U-boats, and what was left of the German surface ships, could still carry out a nuisance raid or deal some last avenging action; so while the wireless and newspapers announced our victories on land, we took up our role of patrolling the North Sea in the winter months of 1944–1945, the last year of war, so we hoped.

At Christmas time I went on the messdeck, for as a Petty Officer it was my duty to see that all was shipshape and the standard maintained. 'No flags to decorate the Mess?', I asked, 'no goodwill amongst shipmates? Where is the joyful spirit of Christmas? It's always the custom to decorate the messdeck and bring out your presents from home. Come, come, my lads,' said I, 'let's cheer up and be grateful. I feel sure for most of you this is going to be your last wartime Christmas, don't let old Jerry get us down. If we haven't got presents of nuts, oranges, chocolates, dates and crackers, you've got photos of your wives, families, sweethearts, girlfriends, even your latest pin-up will do. I will go and get some flags,' and I set about bringing a dash of colour, and with a few items I purchased from the canteen and placed upon the mess tables I had the first one helping me. Another joined in. Soon that messdeck was livening up. Streamers overhead, pictures of mothers, wives, sweethearts and babies amongst the film stars and flimsily dressed maidens appeared from out of the lockers and were placed on the mess table. To capture the goodwill I was hoping to set up a burst of music from a mouth organ. 'Hark the Herald' came floating through the messdeck, taken up and joyously sung, and I realised the spirit was there. Now to keep them going, for I wasn't one talented to put on a turn myself but my heart was touched. There was something I could do.

One privilege of being a Petty Officer was drawing your rum ration neat, whereas the other ratings had it with two-thirds water. Thus I was able to bottle some for bribery or corruption when requiring some favour or privilege that wasn't written in King's Regulations or Admiralty Orders! I silently left, to return a few minutes later with my bottle of 'neaters', which I had been storing to smuggle home on my next home leave. Now, to keep good cheer going, I gave one and all, as far as it would go, a taste of 'Nelson's blood', for it pleased me wholeheartedly that I had helped to restore goodwill and cheerfulness in the place of desultory aloofness.

But all was not over on that Christmas Day! In the evening as I sat playing 'uckers' there was a request at the mess entrance for Petty Officer Sanderson. I stepped into the flat to be confronted by a young A.B. asking if I would present myself on the messdeck. 'I won't keep you a minute and will be right along,' I said as I re-entered the mess, putting on my uniform coat and hat to pay my respects in a proper manner, wondering if there had been some slight trouble in the day's celebration. As I doffed my cap, stepping into the ratings' quarterdeck living compartment, ducking under the flags still displayed by our mornings work, there, grouped around three or four mess tables, were gathered most of the Leading Hands, A.B.s and young, ordinary seamen of my quarterdeck part of the ship. They cordially invited me to come into the middle of the group where a two-badge A.B. waved his hand to indicate silence as I stood rather bashful and curious to see what it was all about.

'We wish to thank you', began the A.B. 'for your help this morning in brightening up this Christmas Day. It is to you we owe the pleasant and enjoyable time we have all had.' There sounded round the gathering 'Hear, hear', followed by three cheers. As the cheers burst forth I realised there wasn't any trouble; I made a short reply, saying something about how I understood, for at times I felt this way myself.

'We wish you to take this gift from us in appreciation of what you did,' carried on the A.B., lifting from the table a sailor hat, holding it in two hands and presenting it to me. Now I became startled with surprise as I stared down at a hat full of money thrust out for me to take – hesitating to grasp the present and for the first time in my life not taking money now offered, as my thoughts jumbled round in my head and I looked down at the heap of money and my hands remained down at my side. At last, somewhat recovering from being taken back, I gasped out 'I am very grateful for what you have done, but something tells me it is not right being a Petty Officer to accept this money. I have a family at home just like you have, and if you write home saying you had a very happy Christmas, this will give satisfaction to all,' I answered. This was accepted by the assembled sailors and I withdrew, disturbed in myself to be so honoured in an unexpected way, which touched me to the heart, yet fearful of the consequences that the favour the rating had displayed to me was not something that one could mention to my fellow petty officers. The kindly act could be embarrassing; it might seem that I was grovelling to those under me, instead of maintaining authority by strict discipline. I decided not to say a word. When I returned to the petty officers' mess and was asked what the call was about I passed it off by saying that an A.B. wanted to see me on a personal request.

A couple of weeks later, I received a letter from my wife saying she had received a sum of money to buy our son a present, and it was signed from the seamen of QX Division in gratitude to his father Petty Officer Sanderson in bringing good will and cheerfulness on a Christmas morning.

CHAPTER THIRTY-TWO

When 1945 came round we were still fighting this war, although the advancing troops had liberated most of occupied Europe and were about to set foot in the almighty Fatherland itself. The Navy now found its task somewhat lightened as it lay waiting, keeping watch over the last remaining waters that the few active German warships could sail, although they still held the long Norwegian coastline. To lie waiting for the end to come is a most dangerous time, for you are apt to lose concentration, becoming careless as you await the victory shouts. A cornered enemy is always liable to make a last attempt to wreak destruction and take life.

So the days slowly ticked by, writing letters home, thinking we could soon put our own private thoughts into words after all the years of censorship. The Navy had changed so much; the large number of new recruits were now taking over from the well-trained sailor, who by now had become scarce.

Perhaps I was getting weary, waiting through those long, cold winter months in Scapa Flow. There was the occasional alarm when we would dash across to Norway, only to return the next day not meeting up with the enemy. We enjoyed a few discussions on what to face up to when we were released from the shackles of the Service, and I think I slowly began to realise my Navy never would be the same any more. It had almost accomplished its task and proved itself the force that the Germans couldn't master, for they only had to win one battle at sea and victory would have been theirs. Now the oceans were almost free from the torpedoing U-boats, aircraft were patrolling thousands of square miles of ocean. Would the ships be required after the conflict or lie rusting like the German hulks still scattering the waters of Scapa Flow?

Less and less were we required to go to sea. Our fighting warship now began to take up a peacetime pattern, chipping and painting, division and church service on Sunday mornings, a film show in the hangar at night, games and competitions arranged with organised discussions on life in a civilian world, on what we wanted after this war was over.

A few lookouts remained, keeping constant watch for aircraft, but now we could sling our hammock almost knowing we would have a night's rest. An exercise or two reminded us all that the war wasn't over. Besides, there was another war on the other side of the world, where we would have to link up with the Americans and finally settle and so I waited thoughtfully, wondering how it was all going to turn out.

'What's going on? Surely we are not going to sea? I'm beginning to settle down, getting used to my old hammock again.' I made this remark to my companion, him being the Senior Petty Officer, for the engine room began to throb with vibration and heated air came roaring out of the air shaft from below.

The loudspeaker close by announced: 'Special sea dutymen close up. Cable party muster on the fo'c's'le. Hands will close up for action stations in ten minutes time.'

'Away goes my night's sleep in the old hammock,' I muttered, as the noise from the engine room increased and smoke belched from the funnels. I hurried up my inspection of the upper deck, carried on down to the Mess, sorted out my lifebelt and necessary clothes to keep myself warm for the night's ordeal of sitting in my director. The loudspeaker announced 'Action Stations'. We turned away to our respective posts.

The Captain announced that enemy ships had been reported keeping close to the Norwegian coast. According to calculations we should intercept the enemy at approximately 0200. This we had heard so many times – and we never did. I often thought the Germans had been alerted also, with both forces avoiding action. Besides, who wanted any action at sea now that the allied forces were closing in? Home sweet home was uppermost in our minds, or certainly in mine, but at that moment we were speeding through the waves bent on our mission, accompanied by another cruiser of our own class. Two young marine officers seated just behind and slightly above me, the Australian commissioned gunner by my side, who pressed the trigger that fired the turrets simultaneously, myself as trainer, all the fire power of the 6″ cruiser, relying on our skill for accuracy. Only the illuminated dial glowed in the dark interior as we settled down, with hardly a word spoken except for our closing-up procedure.

A startled shout 'ALARM! ALARM!' I saw huge splashes shooting up, smothering the *Mauritius* in sheets of spray as I felt her shaking to get up speed, turning to avoid the falling shells. I lined my sight upon that dark shape where the sheet of flame had come from. I shouted out 'trainer on', knowing full well that flash meant a broadside was on its way, directed at us. Long, long seconds of waiting. 'Ding, ding' of the fire gong. *Mauritius* guns thunder out, flames belch forth licking the night. Our shells are on their way. For a split second the jar of the broadside, the flash and belching smoke throw me off my target. I bring my sights back to that target, ready and steady for the next broadside, in which I can detect every movement, the shudder of each salvo, the violent jar, with shaking speed and tilt of slewing round, spouts of water from the falling shells seem to be drenching the *Mauritius*. Could we come through this ferocious and surprise attack, as the German gunfire is accurate and concentrated? Flashes from their broadsides fall in the path of the turning *Mauritius*. Surrounded by the enemy bursts, our guns thunder out in fierce reply to the avalanche of their smothering attack. Keeping calm under the German heavy bombardment plainly I see the dark outline of a German ship speeding across my sights. I keep my vertical wire glued on that target as a black burst envelops the director, knowing we had been hit somewhere below, but the full force of our four triple 6" turrets are roaring their defiance. The shudder of her steel frame as broadside after broadside are sent on their way. The enemy have pressed their attack from 14,000 yards to 3,000 yards. I see a crimson blob of flame and it wasn't a flash from their guns. We must have scored a hit. Now I observe a rolling cloud of belching smoke pouring from the funnels of a speeding vessel. the enemy are putting down a smoke screen, and the target disappears from my binoculars to line-of-sight training on the radar screen and report on by L.S.T., and we carry on the battle although the enemy has disappeared behind their smoke screen. 'Train right,' I hear, hastily switching to my binoculars to see a ship breaking through a cloud of dense smoke to fire its torpedoes. I calmly report 'trainer on' to the excited yelling of the young marine officers. Our guns roar and the daring attacker turns to the safety of the smoke screen. Our shells follow, pitching into that dense pall of blackness, and it appears to me our enemy are playing hide and seek, with me having to change from visual sighting to radar and back again, listening to the excited voices of the young officers to train right or left as they spotted a ship breaking through the smoke screen. And so the battle carried on, but I began to notice the spouts of water from the falling shells are less numerous and beginning to fall some distance away. Gradually we seemed to be getting on top, driving the enemy away; or are they escaping down a creek, luring us into some coastal shore batteries? For that dark shape ahead can't be all smoke, and it sounds to me as if more guns are joining in, and it's our turn now to break off the engagement and turn for the open sea, leaving a zigzag wake, dodging the enemy's last few salvoes. No more firing, and I began slowly to relax from the winding tension of the last hectic hour, to realise we had fought a battle and come out alive and well. But what of the other members? 'We have been hit,' I gasped out. I know by the black burst which was so dense. I couldn't see.'

'They will let us know shortly,' was my companion's reply, as I began to talk, forgetting the difference between rank and file. 'There must have been a German flotilla and we ran into it and they caught us, not us them, or they've got better radar than we have. And where's our escort, the other cruiser?' I had forgotten about it. 'I didn't see it during the action.'

'It's been with us all the time,' was passed on to me.

'I hope there are no casualties,' now taking in more fully what had happened and thinking back, I realised I hadn't been a bit frightened. In all other engagements I had been scared stiff. From our sudden attack I had been cool, calm and collected, even with the pressure of the over-excited young marine officers, who did in a way apologise to me later by saying I had done a splendid job. I realised that the action had come so suddenly and unexpectedly it didn't give me time to think of the consequences, for it was the enemy who had discovered us first and had fully half a minute to get their shell in, but as soon as we got going our steadiness and accuracy overcame the advantage of their surprise attack, for I always had confidence in our supremacy over the Germans not only in seamanship but in our gunnery as well.

With dawn beginning to break a Coastal Command plane came out to meet us, seeing the large gaping hole above the waterline in the fo'c'sle messdeck. Thank goodness no casualties, but alas our sister ship with no scars to show had lost two young sailors hit by flying shrapnel, which struck a note of sadness. To think so near the end of this long war two souls had to depart when the reward of five years of sacrifice was only a few days away.

Ad now the triumphant arrival in Scapa, standing to attention as we passed the mighty battleships anchored in their sheltered waters, never going to sea, displaying our big gaping hole. A pint in the old canteen tonight just to tell those shore-based sailores there was still a war on.

CHAPTER THIRTY-THREE

The *Mauritius* went for a big refit – to be ready for the Far East was the general comment on the lower deck – and I was back in the old barracks at Chatham. There I hoped to await the final overthrow of the almighty Fatherland. The Russians were rushing through the Balkan countries, Italy had surrendered. Country after country was liberated. It became a race to Berlin, and I could await the outcome reading the morning paper.

Again I marched on the parade ground, climbed the steps to the large blocks of Chatham Barracks under the stern look of the figureheads of England's great Admirals so familiar to this establishment yet full of the coming and going in this fateful time, the dockyard ringing out its active repairing and building. Yet somehow there was a calmness in awaiting the end of the war, as if we all resolved this wasn't going to be a sudden stop with millions becoming idle, but a better life ahead, a real and lengthy build-up. We had fought and won and now the social reform, an organised demobilisation was talked of and hinted at by our papers. Not altogether disaster like the last war, with our boys coming home and no work.

Not for me the victory parade or to rejoice in the streets on our final day of deliverance. Away – sent to take passage on a troopship to Malta to join H.M.S. *Atherstone*, a small destroyer. Four days out in the midst of a convoy when the whole country rejoiced in the German collapse. For my celebration there was chicken added to the menu and a small bottle of beer each and on this beverage the lookouts, who I was responsible for, kept reporting U-boats sighted. All I could do to bemoan my fate was to walk up and down, being thankful I was one of the fortunate ones to be spared.

We landed at Malta, to be sent on to Italy, then across Italy to pick up a large tank landing craft, to finally catch up with my old ship at the Adriatic port of Trieste. There it looked as if the divided population of Italians and Yugoslavians were going to kill each other over who should own it, for there were uniformed patrols, seven or eight in number, women included, patrolling the city armed with rifles, automatic guns with bandoliers of hand grenades, so it was a tense and unwelcome time to be amongst this squabbling over the distribution of the spoils of war. A difficult and thankless task lay ahead for our statesmen in the peace settlement to please the small nations in their complicated jealous nationalism. I joined my small destroyer, and when on shore I had to walk past the armed patrols, taking little notice but hoping they wouldn't start throwing their hand grenades while I was ashore. I had a little time seeing what it was like in the beaten countries that this war had left behind. In Naples we entered the port for repairs and it was here I celebrated V.J. Day, in much the same modest way as V.E. Day. The war out East had moved so quickly with atom bombs being dropped and the astonishing surrender of Japan was something unthinkable to us, who had known the fear of being sent out to the Far East, to finally kill these unyielding yellow men to the last man. What bomb so devastating and destructive had scientists invented?

And so my war years had come to an end, and my service for my country had run out. I had to await demobilisation, which had been worked out by age and service, so that in a few weeks' time I would leave for passage home, to be discharged.

Now in my travelling around I had observed that Italian money was almost worthless. In fact small boys had bundles of it, ever ready to buy food or cigarettes in the black market dealing, and you were pestered by these touts. One hundred lire was the official rate of exchange, but anyone would give you 250 for an English pound, so I became a financial wizard. With a few weeks in front of me I might as well go home with a few pounds in my belt! My thirty shillings once a fortnight didn't enable me to spend lavishly so I changed my English pounds on the black market, collecting 250 lire for each, taking it out in savings certificates at a bank in Naples freshly opened for the Forces to save their money. This only remained in business for a few days, for when I went along two days after, it was closed so I had to await my arrival in Malta to present myself at the Paymaster's Office to change my accumulated lire into English pounds. How much did I have when I came out here? How long out and how much had I picked up at the pay table? After these calculations it worked out to be exactly the amount I had to change. 'Don't you ever spend any?', asked the inquisitive officer, paying out reluctantly.

'Oh, no,' I replied, 'don't smoke, don't drink or buy anything for myself. No,' I repeated, adding 'what have I got to spend my money for; as I'm going home shortly I am saving it.'

'So it seems', said the disbelieving officer, paying out reluctantly with a contemptuous look at not finding a flaw in my prepared defence, and when I embarked on a ship I again changed my pounds to lire, for en route home was via Marseilles, across France by train to a transit camp at Calais and then to England, and we would be allowed to change only up to £10 of foreign money, so I had to get the help of a couple of shipmates who hadn't got £10, so possibly I made £100 in my last few weeks of loyal service, a valuable nest-egg for my discharge into civilian life.

I entered Chatham Barracks once again. It had been twelve-and-a-half years earlier that, as a raw country boy, I had stood uncertain before a gold-ringed officer who informed me that I had now officially become an ordinary seaman in the Royal Navy, so proud and pleased that I had made it. As a Petty Officer first-class gunnery rate of good conduct, with almost six years of the twelve having been in active war operations, returning for always to my loving wife to rejoin the village life after my worldwide adventures. perhaps inwardly I still had that boyhood ambition to have 'a hundred pounds and a farm'. I had the money, and attached to my father's pub was land to be cultivated. I carried out my demobilisation routine in the next few days, to be given my railway ticket, bag and hammock and transported on a lorry to Chatham Station. No gold-ring officer to say farewell. Sadly, it seemed I just faded out of the service with my bag and faithful old hammock.

PART OF THE LIFE STORY OF R. D. SANDERSON